METHODS IN MOLECULAR BIOLOGY

For further volumes:
http://www.springer.com/series/7651

For over 35 years, biological scientists have come to rely on the research protocols and methodologies in the critically acclaimed Methods in Molecular Biology series. The series was the first to introduce the step-by-step protocols approach that has become the standard in all biomedical protocol publishing. Each protocol is provided in readily-reproducible step-by step fashion, opening with an introductory overview, a list of the materials and reagents needed to complete the experiment, and followed by a detailed procedure that is supported with a helpful notes section offering tips and tricks of the trade as well as troubleshooting advice. These hallmark features were introduced by series editor Dr. John Walker and constitute the key ingredient in each and every volume of the Methods in Molecular Biology series. Tested and trusted, comprehensive and reliable, all protocols from the series are indexed in PubMed.

Fusarium wilt

Methods and Protocols

Edited by

Jeffrey Coleman

Department of Entomology and Plant Pathology, Auburn University, Auburn, AL, USA

Humana Press

Editor
Jeffrey Coleman
Department of Entomology and Plant
Pathology
Auburn University
Auburn, AL, USA

ISSN 1064-3745 ISSN 1940-6029 (electronic)
Methods in Molecular Biology
ISBN 978-1-0716-1794-6 ISBN 978-1-0716-1795-3 (eBook)
https://doi.org/10.1007/978-1-0716-1795-3

This Humana imprint is published by the registered company Springer Science+Business Media, LLC part of Springer Nature.
The registered company address is: 1 New York Plaza, New York, NY 10004, U.S.A.

Preface

Fusarium wilt diseases are primarily caused by members of the *Fusarium oxysporum* species complex (FOSC) and are found worldwide. These filamentous phytopathogenic fungi are able to infect over 120 genera of plants, including tomato, banana, cucurbits, melon, cotton, and a wide range of fruits and vegetables. In addition to their broad host range, these fungi have many interesting biological aspects that have been the focus of investigation; in particular they have been well established to harbor additional chromosomes (i.e., the accessory genome) that confer many isolate-specific traits, including host specificity and niche adaptation. The unique characteristics of members of the FOSC have prompted researchers to develop and optimize several molecular procedures to study their biology. In this volume, a collection of molecular protocols detailing some of the most common and modern techniques for use with these fungi is presented. These methods range from the initial isolation and molecular-based identification to further specific protocols involving genome characterization, generation of mutants, and characterization of interactions with other organisms including host plants. This volume could serve as a valuable laboratory resource for mycologists, plant pathologists, microbiologists, geneticists, and others that have an interest in the FOSC or closely related fungi.

Auburn, AL, USA *Jeffrey Coleman*

Contents

Contributors

Yu Ayukawa • *Center for Sustainable Resource Science, RIKEN, Yokohama, Japan*
Ernesto Pérez Benito • *Instituto Hispano-Luso de Investigaciones Agrarias (CIALE), Dpto. de Microbiología y Genética, Universidad de Salamanca, Salamanca, Spain*
Virginia Casado-del Castillo • *Instituto Hispano-Luso de Investigaciones Agrarias (CIALE), Dpto. de Microbiología y Genética, Universidad de Salamanca, Salamanca, Spain*
Catherine Danmaigona Clement • *Department of Biochemistry and Biophysics, Texas A&M University, College Station, TX, USA*
Jeffrey J. Coleman • *Department of Entomology and Plant Pathology, Auburn University, Auburn, AL, USA*
Kirk J. Czymmek • *Advanced Bioimaging Laboratory, Donald Danforth Plant Science Center, Saint Louis, MO, USA*
Braham Dhillon • *Department of Plant Pathology, University of Florida, Fort Lauderdale Research and Education Center, Davie, FL, USA*
Antonio Di Pietro • *Departamento de Genética, Campus de Excelencia Internacional Agroalimentario ceiA3, Universidad de Córdoba, Córdoba, Spain*
Cristina López Díaz • *Departamento de Genética, Universidad de Córdoba, Córdoba, Spain*
José María Díaz-Mínguez • *Instituto Hispano-Luso de Investigaciones Agrarias (CIALE), Dpto. de Microbiología y Genética, Universidad de Salamanca, Salamanca, Spain*
Yan-Hong Dong • *College of Plant Protection, Hebei Agricultural University, Baoding, China*
Tânia R. Fernandes • *Departamento de Genética, Campus de Excelencia Internacional Agroalimentario ceiA3, Universidad de Córdoba, Córdoba, Spain; GreenUPorto, Sustainable Agrifood Production Research Centre/Inov4Agro & DGAOT, Faculty of Sciences, University of Porto, Porto, Portugal*
David M. Geiser • *Department of Plant Pathology and Environmental Microbiology, The Pennsylvania State University, University Park, PA, USA*
Navdeep Gill • *Department of Crop, Soil and Environmental Sciences, University of Arkansas, Fayetteville, AR, USA*
Lucía Gómez-Gil • *Departamento de Genética, Universidad de Córdoba, Córdoba, Spain*
Ping He • *Department of Biochemistry and Biophysics, Texas A&M University, College Station, TX, USA*
Seogchan Kang • *Department of Plant Pathology & Environmental Microbiology, The Pennsylvania State University, University Park, PA, USA*
Pei-Lun Kao • *Department of Biochemistry and Molecular Biology, University of Massachusetts Amherst, Amherst, MA, USA*
John C. Kennell • *Department of Biology, Saint Louis University, Saint Louis, MO, USA*
Hye-Seon Kim • *Mycotoxin Prevention and Applied Microbiology Research Unit, USDA, Agricultural Research Service, National Center for Agricultural Utilization Research, Peoria, IL, USA*

Imana Laraba • *Mycotoxin Prevention and Applied Microbiology (MPM) Research Unit, ARS-USDA, Peoria, IL, USA*

Ningxiao Li • *Department of Plant Pathology & Environmental Microbiology, The Pennsylvania State University, University Park, PA, USA*

Jinggao Liu • *Southern Plains Agricultural Research Center, Agricultural Research Service, USDA, College Station, TX, USA*

Zunyong Liu • *Department of Biochemistry and Biophysics, Texas A&M University, College Station, TX, USA*

Li-Jun Ma • *Department of Biochemistry and Molecular Biology, University of Massachusetts Amherst, Amherst, MA, USA*

Michael C. MacKillop • *Department of Biology, Saint Louis University, Saint Louis, MO, USA*

Sibongile Mafu • *Department of Biochemistry and Molecular Biology, University of Massachusetts Amherst, Amherst, MA, USA*

Eun-Gyu No • *Department of Biochemistry and Biophysics, Texas A&M University, College Station, TX, USA*

Kerry O'Donnell • *Mycotoxin Prevention and Applied Microbiology (MPM) Research Unit, ARS-USDA, Peoria, IL, USA*

Sook-Young Park • *Department of Plant Medicine, College of Life Science and Natural Resources, Sunchon National University, Suncheon, South Korea*

Elena Pérez-Nadales • *Departamento de Genética, Universidad de Córdoba, Córdoba, Spain; Maimonides Biomedical Research Institute of Cordoba (IMIBIC), Reina Sofia University Hospital, Universidad de Córdoba, Córdoba, Spain*

Ambika Pokhrel • *Department of Entomology and Plant Pathology, Auburn University, Auburn, AL, USA*

Akaansha Rampal • *Department of Biochemistry and Molecular Biology, University of Massachusetts Amherst, Amherst, MA, USA*

Sergey Savinov • *Department of Biochemistry and Molecular Biology, University of Massachusetts Amherst, Amherst, MA, USA; Department of Chemistry, Tufts University, Medford, MA, USA*

Seungyeon Seo • *Department of Entomology and Plant Pathology, Auburn University, Auburn, AL, USA*

Antonio Serrano • *Departamento de Genética, Campus de Excelencia Internacional Agroalimentario ceiA3, Universidad de Córdoba, Córdoba, Spain; Université Côte D'Azur, CNRS, INSERM, Institute of Biology Valrose (iBV), Nice, France*

Libo Shan • *Department of Biochemistry and Biophysics, Texas A&M University, College Station, TX, USA*

Masatoki Taga • *Department of Biology, Faculty of Science, Okayama University, Okayama, Japan*

Gesabel Navarro Velasco • *Departamento de Genética, Universidad de Córdoba, Córdoba, Spain; LEAF-Linking Landscape, Environment, Agriculture and Food, School of Agriculture, University of Lisbon, Lisbon, Portugal*

Tanya A. Wagner • *Southern Plains Agricultural Research Center, Agricultural Research Service, USDA, College Station, TX, USA; Department of Plant Pathology and Microbiology, Texas A&M University, College Station, TX, USA*

QIANG WANG • *State Key Laboratory of Crop Stress Biology for Arid Areas and College of Plant Protection, Northwest A&F University, Yangling, Shaanxi, China*
SHU-TONG WANG • *College of Plant Protection, Hebei Agricultural University, Baoding, China*
CATHERINE C. WASMANN • *The School of Plant Sciences, University of Arizona, Tucson, AZ, USA*
YONG ZHANG • *Department of Biochemistry and Molecular Biology, University of Massachusetts Amherst, Amherst, MA, USA; Bioinformatics Section, National Institute of Neurological Disorders and Stroke, National Institute of Health, Bethesda, MD, USA*

Chapter 1

Pure Culture and DNA Sequence-Based Identification of *Fusarium* from Symptomatic Plants and Diverse Substrates

Kerry O'Donnell, Imana Laraba, and David M. Geiser

Abstract

Fusarium ranks as the most important group of plant pathogens, responsible for a wide range of economically destructive diseases, including vascular wilts and root, crown, and stem rots. In addition, head blight and ear rot diseases are associated with the accumulation of mycotoxins in cereals. With over 300 phylogenetically distinct species, and a dearth of phenotypical characteristics, DNA sequence data in most instances is the only reliable means for obtaining an accurate species identification. Here we describe how to obtain single-spored pure cultures from symptomatic host tissue and a molecular identification by querying publicly accessible DNA sequence databases using a portion of translation elongation factor 1-α (*TEF1*), the largest subunit of RNA polymerase (*RPB1*), and/or the second largest subunit of RNA polymerase (*RPB2*).

Key words ABI sequence chromatogram, BLASTn query, DNA extraction, DNA sequencing, FUSARIUM-ID, *Fusarium MLST*, NCBI GenBank, PCR amplification, Pure culture

1 Introduction

Phytopathologists, clinical microbiologists, and other diverse scientists who encounter *Fusarium* require robust methods for accurately identifying the etiological agent to optimize disease management and infection control [1, 2]. Over the past two decades, genealogical concordance among portions of multiple phylogenetically informative genes (GCPSR sensu [3]) has become the gold standard for species recognition and identification in *Fusarium* and virtually all other fungi where this approach has been applied. GCPSR-based studies on *Fusarium* have revealed that this genus comprises 5–10 times more phylospecies than represented in identification guides that employed morphological species recognition [4–6], with new species being discovered at a high rate. Portions of three orthologous genes have been employed in

Jeffrey Coleman (ed.), *Fusarium wilt: Methods and Protocols*,
Methods in Molecular Biology, vol. 2391, https://doi.org/10.1007/978-1-0716-1795-3_1,

many of these molecular phylogenetic studies because they resolve at or near the species level and exist as single copy orthologues throughout the genus. These include translation elongation factor 1-α (*TEF1* [7]) and DNA-directed RNA polymerase II largest (*RPB1*) and second largest subunit (*RPB2*). Conserved oligonucleotide primers have been designed that make it possible to PCR amplify and Sanger sequence these three loci across the breadth of the genus [8, 9]. Sequence chromatograms obtained with an automated DNA sequencer such as an Applied Biosystems DNA analyzer must be carefully edited with software (e.g., Sequencher, Gene Codes Corp., Ann Arbor, MI) to make sure they are error-free before using them to query a web-accessible DNA sequence database. Once chromatograms are error-free, the sequences can be used to conduct BLASTn queries of FUSARIUM-ID (http://isolate.fusariumdb.org/; [10]), *Fusarium MLST* (http://fusarium.mycobank.org/; [8]), and/or NCBI GenBank (https://www.ncbi.nlm.nih.gov/nucleotide). Because many *Fusarium* sequences in NCBI GenBank are misidentified [11], BLASTn queries of it should always be done together with FUSARIUM-ID or *Fusarium MLST* to help compare and interpret the results. Although sequence from one of the three loci, especially *TEF1*, can often be used to obtain a definitive species level identification, similar queries using *RPB1* or *RPB2* are recommended to obtain independent confirmation of the identification. Comparative genomics is providing a robust means for identifying additional phylogenetically informative orthologues for species level studies and effector genes for typing strains for host specificity in the *F. oxysporum* species complex [12].

2 Materials

Personal protective equipment (PPE), including a buttoned lab coat, protective gloves, and safety goggles, should be worn at all time when handling hazardous reagents such as ethidium bromide (EtBr), acetic acid, and sodium hydroxide. Many *Fusarium* species, including some plant pathogens, cause rare human infections, mostly in immunocompromised individuals. Most of these species are normally common in the human environment. As such, researchers should handle cultures minimally using CDC Biosafety Level 1 standards (https://www.cdc.gov/cpr/infographics/00_docs/biosafety.pdf), but additional Level 2 standards, particularly use of a Class II biosafety cabinet, may be required or recommended. Individual workers should follow regulations and guidelines of their institution, as well as USDA-APHIS Plant Protection and Quarantine permit conditions. Bottles containing media/reagents that are placed in a microwave or autoclave must be carefully inspected to make sure the caps are loose. Caps on media/reagents that have been heated should never be tightened until the solutions have cooled to room temperature. All culture media should be prepared with distilled or Type II H_2O. Sterilized agar media is stored at room temperature in L bottles with the cap tightly closed once the agar has cooled. Stock solutions for DNA extraction, the polymerase chain reaction (PCR), and DNA sequencing should be prepared with molecular biology grade reagents and sterile molecular grade H_2O. Whether a stock solution is stored at room temperature or −20/−80 °C is indicated below for each reagent. Regulations regarding waste disposal should be strictly adhered to.

2.1 Isolation and Pure Culture of Strains from Symptomatic Hosts

1. Commercial bleach (5.25–8.25% sodium hypochlorite): dilute 1:9 in distilled H_2O to obtain a 0.5–0.8% working solution (*see* **Note 1**).
2. 3% H_2O agar with and without antibiotics (sterile): add 15 g microbiological agar to 500 mL distilled H_2O in a 1 L bottle with cap loosened, and microwave until agar is dissolved. If necessary, swirl gently to make sure all of the agar is dissolved. Autoclave for 15 min at 121 °C and 15 psi. Place in a 55 °C water bath until cooled. For 3% H_2O agar with antibiotics, in a biosafety hood add 10 mL penicillin (~5000 units)/streptomycin (5 mg)/neomycin (10 mg) per mL (PSN) solution to the molten 3% H_2O agar (*see* **Note 2**), whirl to mix, and then pour ~5-mm-thick layer in 60 × 15 mm plastic disposable Petri dishes. Leave the plates in the biosafety hood at room temperature. When plates have completely solidified, wrap them with cellophane, and store in a 4 °C refrigerator (*see* **Note 3**). Follow the same procedure to prepare 3% H_2O agar plates without antibiotics, but place a 1 cm^2 sterile Whatman grade 1 filter paper in the center to promote conidiation of single-spored pure cultures.
3. Yeast-malt broth: add 3 g yeast extract, 3 g malt extract, 5 g peptone, 20 g glucose to 1 L of distilled H_2O, and stir and microwave if necessary, until dissolved. Dispense 500 mL into two 1 L bottles, and autoclave for 15 min at 121 °C and 15 psi with caps loosened (*see* **Note 4**). Once autoclaved and the media has cooled, tighten the caps and store at room temperature.
4. 15% glycerol (sterile): pour 150 mL glycerol into 1 L graduated cylinder, and bring volume to 1 L with type II H_2O. Tightly cap cylinder top with parafilm, firmly press hand over top to prevent leakage, and gently invert several times until fully mixed. Dispense ~80 mL into 160 mL bottles, and autoclave for 15 min at 121 °C and 15 psi with caps loosened (*see* **Note 5**). Once sterilized glycerol has cooled, tighten caps and store at room temperature.

2.2 Culturing Strains for Isolation of Total Genomic DNA

1. 50 mL disposable polypropylene centrifuge tubes.
2. Büchner funnel attached to 2 L Erlenmeyer flask, with inline trap flask and house vacuum (*see* **Note 6**).
3. Freeze dryer (*see* **Note 7**).
4. CTAB buffer: 100 mM Tris-Cl pH 8.4, 1.4 M NaCl, 25 mM EDTA, 2% hexadecyltrimethyl ammonium bromide (CTAB). To prepare 0.5 L CTAB buffer, add 50 mL of 1 M Tris-Cl pH 8.4, 40.9 g NaCl, 25 mL 0.5 M EDTA pH 8.0, and 10 g CTAB to a 1 L beaker, bring volume up to 0.5 L with molecular grade H_2O, and dissolve by heating in a 55 °C water bath (*see* **Note 6**). In a chemical fume hood, prepare 1 M Tris-HCl pH 8.4 by dissolving 121 g of Tris base in 800 mL molecular grade water in a 1 L glass beaker containing a magnetic stir bar on a stir plate. Insert a pH meter in solution and adjust to a pH of 8.4 with ~32 mL of concentrated HCl, and then bring the volume up to 1 L and mix thoroughly. Prepare 1 L 0.5 M EDTA by adding 186.1 g disodium ethylenediaminetetraacetate dihydrate (EDTA) to a beaker containing 800 mL of H_2O and a magnetic stir bar, and stir vigorously on a magnetic stir plate. Using a pH meter, adjust the pH to 8.0 with NaOH by gradually adding 18–20 g of NaOH pellets (*see* **Note 7**). Once fully dissolved, bring the volume up to 1 L with molecular grade H_2O and mix thoroughly.
5. 100–1000 μL single channel micropipette.
6. Molecular biology grade chloroform.
7. 70% ethanol (ETOH).
8. Isopropanol.
9. Microfuge with a rotor holding 24–48 1.5 mL microfuge tubes.
10. Digital dry block heaters.
11. Molecular grade pure water.

2.3 PCR Amplification, DNA Sequencing, and Strain Identification

1. 96-well plates for PCR and sequencing (*see* **Note 8**).
2. 0.1–2 μL, 2–20 μL, 20–200 μL, 100–1000 μL single channel micropipettes; 8 channel 20–200 μL and 12 channel 2–20 μL micropipettes; and filter pipette tips (*see* **Note 9**).
3. 2× PCR master mix (sterile; *see* **Note 10**): in 1.5 mL microfuge tube, combine 560 μL sterile molecular grade H_2O, 200 μL 10× PCR buffer, 100 μL 25 mM $MgSO_4$, 40 μL deoxynucleotide triphosphate mix (10 pmol/μL of each dNTP) mix (*see* **Note 11**), 50 μL forward and reverse PCR primers at 10 pmol/μL (*see* **Note 12**; Table 1). See **steps 1–3** in section 3.3 of the Methods for instructions on how to use the 2× PCR master mix.

Table 1
Primers used to amplify loci of interest

Locus	Gene product	Amplicon size (bp)[a]	Primer	Sequence (5′–3′)[b]	PCR[c]	Sequencing
TEF1	Translation elongation factor 1-α	639–683	EF1	ATGGGTAAGGA**R**GACAAGAC	•	
			EF2	GGA**R**GTACCAGT**S**ATCATG	•	
			EF3	GTAAGGAGGA**S**AAGACTCACC		•
			EF22i	AGGAACCCTT**H**CCGAGCTC		•
RPB1	RNA polymerase largest subunit	1607	Fa	CA**Y**AA**R**GA**R**TC**Y**ATGATGGG**W**C	•	•
			G2R	GTCAT**Y**TG**D**GT**D**GC**D**GG**Y**TC**D**CC	•	•
			F5	ATGGGTAT**Y**GTCCAGGA**Y**TC		•
			F7	C**R**ACACAGAAGAGTTTGAAGG		•
RPB2	RNA polymerase second largest subunit	1700–1742	5f2	GGGG**W**GA**Y**CAGAAGAAGGC	•	•
			7cr	CCCAT**R**GCTTG**Y**TT**R**CCCAT	•	•
			7cf	ATGGG**Y**AA**R**CAAGC**Y**ATGGG	•	•
			11ar	GC**R**TGGATCTT**R**TC**R**TC**S**ACC	•	•

[a]*RPB2* is amplified as two contiguous fragments with 5f2 × 7cr and 7cf × 11ar

[b]Degenerate sites in bold indicate the following: D = A or G or T; H = A or C or T; R = A or G; S = G or C; W = A or T; Y = C or T

[c]EF1 × EF2 Tm = 52 °C; Fa × G2R Tm = 55 °C; 5f2 × 7cr and 7cf × 11ar Tm = 55°

4. 10× PCR buffer (sterile): 200 mM Tris-HCl (pH 8.4), 500 mM KCl. To a 1.5 mL microfuge tube, add 200 μL 1 M Tris-HCl (pH 8.4), 500 μL 1 M KCl, and 300 μL molecular grade H_2O. Vortex to mix and store at −20 °C.
5. 1 M KCL (sterile): Add 7.4 g KCl to 100 mL graduated cylinder, and bring volume up to 100 mL with molecular grade H_2O and mix to dissolve. Sterilize by autoclaving in 160 mL milk dilution bottle with cap loosened.
6. 1 M $MgSO_4$ (sterile): Add 24.6 g of $MgSO_4$•$7H_2O$ to a 100 mL graduated cylinder, and bring volume up to 100 mL

with molecular grade H_2O. Stir to dissolve. Sterilize by autoclaving in 160 mL milk dilution bottle with loose cap. Prepare 25 mM stock for PCR by diluting 25 μL 1 M $MgSO_4$ in 975 μL sterile molecular grade H_2O in a cryotube, and store at −20 °C.

7. dNTP mix (sterile): Thaw 100 mM dATP, dCTP, dGTP, and dTTP Na-salt, PCR grade deoxynucleoside triphosphate set, vortex, and centrifuge briefly. To a 1.5 mL microfuge tube, add 100 μL of each dNTP, and bring the volume up to 1 mL with 600 μL of sterile molecular grade H_2O. Vortex to mix and store at −20 °C.
8. PCR primers for *TEF1*, *RPB1*, *RPB2* (Table 1).
9. Thermocycler.
10. Wide horizontal agarose gel electrophoresis apparatus with power supply connected to a timer, agarose gel casting tray, comb with 1-mm-thick teeth, and leveling bubble.
11. 1.5% agarose gel in 1× TAE buffer: To a 500 mL screw-capped bottle, add 300 mL 1× TAE buffer and 4.5 g molecular biology grade ultrapure agarose. Loosen cap and dissolve by microwaving. Tighten cap once agar has solidified. Prepare 1× TAE buffer stock by diluting 20 mL of 50× TAE in 980 mL of molecular grade H_2O.
12. 50× TAE buffer: Wearing PPE, in a chemical fume hood dissolve 242 g of Tris-base (MW = 121.14 g) in 700 mL molecular grade H_2O. Carefully add 57.1 mL 100% acetic acid and 100 mL of 0.5 M EDTA (pH 8.0). Bring the volume up to 1 L with molecular grade H_2O, mix until Tris-base has dissolved on a magnetic stir plate, and store at room temperature with cap tightened.
13. 6× gel loading dye: Add 25 mg bromophenol blue (BPB) to 6.7 mL molecular grade H_2O and mix. Then add 3.3 mL glycerol and mix until BPB is fully dissolved; for long-term storage, aliquot 1 mL into 1.5–2.0 mL cryotubes, and store tightly capped at −20 °C (*see* **Note 13**).
14. 10 mg/mL ethidium bromide (EtBr) stock solution: Wearing PPE, carefully add 1 g of EtBr to 100 mL molecular grade H_2O, and stir on a magnetic stir plate until completely dissolved; or use SYBR Green I nucleic acid gel stain (Sigma-Aldrich) (*see* **Note 14**).
15. Plastic container: Large enough to hold the largest agarose gel and molecular grade H_2O containing 1 drop EtBr/100 mL.
16. Plastic tray for destaining gel in distilled H_2O.
17. UV transilluminator and photo-documentation system.
18. ExoSAP-IT™ stored at −20 °C (*see* **Note 15**).

19. BigDye Terminator™ v3.1 supplied with 5× sequencing buffer and **BigDye XTerminator™** Purification Kit both stored at −20 °C for use with an ABI DNA Analyzer (*see* **Note 16**).
20. MicroAmp™ clear adhesive film (ThermoFisher).
21. Software for correcting sequence chromatograms (*see* **Note 17**).
22. Strain identification via BLASTn queries of publicly accessible DNA sequence databases (*see* **Note 18**).
23. A batch BLASTn query of NCBI GenBank can be conducted if sequences of multiple isolates are being identified.

3 Methods

3.1 Isolation and Pure Culture of Strains from Symptomatic Hosts

1. Wash symptomatic tissues with 3 or more exchanges of sterile distilled H_2O, excise ~5 mm pieces, and surface disinfect them in a 2% bleach (NaOCl) solution for 1 min followed by 3 washes in sterile distilled H_2O for 5 min. In a biosafety cabinet, blot tissue dry on sterile paper towels, and then place them on 3% H_2O agar plates supplemented with 10 mL/500 mL penicillin-streptomycin-neomycin (PSN) solution.
2. Incubate at 25 °C under 12 h white fluorescent and near UV blue-black-blue lights and 12 h dark cycle until conidiating *Fusarium*-like colonies are visible. Under a dissecting microscope, use sterile fine-pointed needle [ex., hypodermic] to excise part of the colony producing microconidia and/or multiseptate, fusiform macroconidia, and transfer to a 1.5 mL microfuge tube containing 1 mL of sterile distilled H_2O.
3. In a biohood, vortex briefly and use a 2–20 μL pipette to transfer 10 μL to a 3″ × 1″ × 1 mm microscope slide and cover with a 22 × 22 mm cover slip. Examine with a compound microscope, and if the conidium concentration is too concentrated to count, serially dilute with sterile distilled H_2O so that there are fewer than 100 conidia/10 μL.
4. Once the conidium concentration per μL of a homogenous mixture has been estimated, plate 10–20 conidia on a 60 mm Petri plate containing 3% H_2O agar with PSN antibiotics, and use a sterile L-shaped microbiological spreader to spread conidia evenly over agar surface. If the agar surface is dry, place 2–3 drops of sterile molecular grade H_2O in the center of the plate, and add conidial suspension to it. Then spread with sterile L-shaped microbiological spreader.
5. After incubating for 24–48 h at 25 °C, examine the plates under a dissecting microscope, and use a sterile fine-pointed

needle to transfer a germling to a new 60 mm Petri plate containing 3% H_2O agar with PSN antibiotics. Once colonies have grown out far enough to remove hyphal tips, in a biohood use a sterile individually wrapped 9″ Pasteur pipette to remove a plug from the colony margin, and eject it into a 50 mL disposable polypropylene centrifuge tube or 125 mL sterile Erlenmeyer flask containing 20 mL of yeast-malt broth. When the broth culture is in log phase, most cultures will appear clouded because, except for *Fusarium graminearum* and related species, many *Fusarium* species produce abundant microconidia in submerged culture. Transfer 10 μL to a microscope slide in a biohood, and examine with a 40× objective in a compound microscope to make sure the culture is not contaminated with bacteria. If the culture is pure, in a biohood use a sterile 9″ Pasteur pipette to transfer several drops of the broth culture to a 2 mL cryotube containing 1 mL of 15% glycerol, vortex briefly, and then store in a −80 °C freezer.

3.2 Culturing Strains for Isolation of Total Genomic DNA

1. In biohood, inoculate a 50 mL disposable polypropylene centrifuge tube containing 20 mL of yeast-malt broth with a plug from the colony margin or spore suspension of a single-spored strain (*see* **Note 19**).
2. Culture at 25 °C in the dark on rotary shaker set at 200 rpm until ample cellular material for DNA extraction is obtained (usually 2–4 days).
3. Harvest mycelium by filtration with a 90 mm Whatman #1 filter paper disc on a 100-mm-diameter Büchner funnel attached to a 2 L Erlenmeyer flask with inline flask trap and house vacuum (*see* **Note 20**).
4. Scrape mycelium off filter paper, place in a labeled 1.5 mL microfuge tube, and attach a Kimwipe to the cap with a rubber band. Store in a −20 °C freezer for 1 h before placing in a freeze dryer for 12–24 h (*see* **Note 21**).
5. Add 50–100 mg of freeze-dried mycelium to a labelled 1.5 mL microfuge tube, and pulverize with a 1 mL disposable pipette tip.
6. Add 700 μL CTAB buffer to each tube and vortex to resuspend pulverized mycelium. Use a 20–200 μL disposable pipette tip to dislodge any unwetted mycelial pellet at the bottom of the tube.
7. In a chemical fume hood, add 700 μL of chloroform to each tube, and shake briefly until an emulsion is obtained (*see* **Note 22**).
8. Spin in a microfuge at 10,464 × *g* for 10 min.
9. In chemical hood, transfer 350 μL of the upper CTAB aqueous phase to a new labelled 1.5 microfuge tube.

10. Precipitate DNA with 350 μL of −20 °C isopropanol and mix by inverting several times.
11. Spin in a microfuge at 10,464 × *g* for 3 min.
12. Gently pour off supernatant, wash pellet with 500 μL 70% ETOH, and spin in a microfuge at 12,000 rpm for 3 min. Gently discard the supernatant and then spin in microfuge for 10 s. Gently flip ETOH out of tube, and then blot lip of inverted tube with a Kimwipe to remove residual ETOH. Resuspend pellet in 100 μL sterile molecular grade H_2O, and store in a −20 °C freezer until ready to use (*see* **Note 23**).

3.3 PCR Amplification, DNA Sequencing, and Strain Identification via BLASTn Queries of Publicly Accessible DNA Sequence Databases

1. Prepare dilute genomic DNA for PCR by adding 4 μL of each genomic DNA to 200 μL molecular grade H_2O in a 96-well PCR plate.
2. Prepare 1 mL PCR master mix for each primer pair (Table 1) *without* Invitrogen Platinum® *Taq* DNA polymerase as follows: 560 μL molecular grade H_2O, 200 μL 10× PCR buffer, 100 μL 25 mM $MgSO_4$, 40 μL dNTP mix, 50 μL forward PCR primer at 10 pmol/μL, and 50 μL reverse primer at 10 pmol/μ L. Vortex and store in −20 °C freezer when not in use.
3. If the total PCR reactions are 30 μL, multiply the number of samples by 15 plus several more to compensate for pipetting errors. Thus, for 20 samples, you will need ~22 × 15 μL = 330 μL of 2× PCR reaction mix +2.64 μL Invitrogen Platinum® *Taq* (i.e., 0.12 μL *Taq*/15 μL 2× master mix), vortex, and spin down briefly. Dispense 15 μL of the 2× PCR master mix + *Taq* to each well of a 96-well plate, or part of it, or individual 200 μL tubes, and then add 15 μL of each dilute genomic DNA.
4. Use the following PCR amplification program: 94 °C for 90 s; 40 cycles of 94 °C for 30 s, 30 s at 52 or 55 °C (depending on the melting temperature of the primers used; Table 1), and 3 min at 68 °C; 5 min extension at 68 °C, followed by a 4 °C hold.
5. Confirm amplicons by running ~2 μL of each PCR reaction mixture and a molecular weight marker into a 1.5% agarose gel in 1× TAE buffer for 18 min (*see* **Note 24**), followed by staining for 15 min in EtBr or Cyber green (*see* **Note 25**). Then destain gel in distilled H_2O for 15 min followed by visualization and photo-documentation over a UV transilluminator (*see* **Note 26**).
6. Prior to sequencing amplicons, primers and nucleotides are removed by adding 4 μL ExoSAP-IT™ reagent to 10 μL of the amplicon. After this mix is incubated at 37 °C in a thermocycler for 30 min, it is heated at 80 °C for 15 min to inactivate the two enzymes in ExoSAP-IT, and then the purified amplicon is sequenced directly.

7. 10 μL of the sequencing reaction mixtures are prepared by combining 4.5 μL of sterile molecular grade H_2O, 2 μL of BigDye sequencing buffer, 1 μL of the sequencing primer, 0.5 μL of BigDye Terminator v3.1, and 2 μL of the purified amplicon. Always prepare a master mix with several extra reactions to compensate for pipetting errors (*see* **Note 27**).
8. Use the following cycle sequencing program: 15 cycles of 10 s at 96 °C, 5 s at 50 °C, 75 s at 60 °C; followed by 5 cycles of 10 s at 96 °C, 5 s at 50 °C, 90 s at 60 °C; followed by 5 cycles of 10 s at 96 °C, 5 s at 50 °C, 2 min at 60 °C; ending in a 4 °C hold.
9. A protocol for using an ABI BigDye XTerminator (BDX) kit (ThermoFisher) to clean up sequencing reactions before they are run on an automated ABI DNA sequencer is provided, but most researchers will have this conducted by a DNA core facility. 40 μL of the following master mix is needed for each 10 μL sequencing reaction: 5 μL BDX, 22.5 μL SAM™ solution, and 12.5 μL deionized water. Vortex to thoroughly mix and then use a wide bore pipette tip to add 40μL of the master mix to each sequence reaction and add 40 μL water to any empty wells. Seal the plate with MicroAmp™ clear adhesive film, and heat in a thermocycler at 160 °C for 5 s. Once completed, vortex the plate for 30 min, and then spin down for 2 min at 1000 g in a centrifuge fitted with a swinging bucket rotor. Samples are stable at room temperature for 48 h and frozen for 10 days. When samples are ready to run on the ABI DNA sequencer, remove adhesive film and replace with septum (*see* **Note 28**).
10. Import sequence chromatograms into a suitable program with a trace viewer such as Sequencher (http://www.genecodes.com/) or Geneious (https://www.geneious.com/) where they are corrected, poor quality data at the ends are removed, and they are assembled as a single contig that is exported as a fasta text file for conducting BLASTn queries of FUSARIUM-ID (http://isolate.fusariumdb.org/blast.php), *Fusarium MLST* (http://fusarium.mycobank.org/), and/or NCBI GenBank (https://www.ncbi.nlm.nih.gov/nucleotide/) (*see* **Note 29**).

4 Notes

1. The length of time host material is surface disinfected in NaOCl will need to be determined empirically, but 1 min is a good starting point. For reproducible results, dilute NaOCl should be made up fresh. Also note that the percent of sodium hypochlorite in commercial bleach varies in concentration (e.g., 5.25–8.25%).

2. The penicillin-streptomycin-neomycin (PSN) solution is commercially available through Sigma-Aldrich, but many other products are available.
3. PSN containing agar plates are good for over 1 month when wrapped in cellophane and stored at 4 °C. If water condensate runs onto the agar, the lid should be removed in a biohood, and the agar surface should be dried before using.
4. Once the yeast-malt broth has cooled to room temperature, tightly cap each bottle to prevent contamination and for long-term storage.
5. We use 160 mL narrow mouth milk dilution bottles and tighten the caps to prevent contamination once the 15% glycerol has cooled to room temperature after autoclaving.
6. Note that there are numerous recipes for preparing CTAB buffer for DNA extraction; however, we have used this one for over 2 decades with uniform success.
7. The last of the NaOH should be added very slowly to help ensure the final pH of 0.5 M EDTA is 8.0. Note that as the pH nears 8.0, EDTA will slowly go into solution.
8. The PCR plate must be compatible with the thermocycler in use. For PCR, we use part or all of a non-skirted 96-well plate covered with a rubber mat in an ABI 9700 cycler and a plate covered with adhesive film in an ABI ProFlex 2 × 96-well PCR System. However, we use skirted 96-well plates for DNA sequencing because they are required for running DNA sequencing reactions mixtures on an ABI 3730 DNA analyzer. Individual tubes or strips of 8 0.2 mL tubes are available if small numbers of samples are PCR amplified.
9. It is essential that labs conducting PCR and DNA sequencing experiments maintain separate clean and PCR product contaminated areas and a set of micropipettes that are dedicated for each. Single and multichannel micropipettes and compatible filter pipette tips in the clean lab area dedicated to DNA isolation, setting up PCR reactions, and DNA sequencing reactions without amplicons should never leave the clean area and never be used for pipetting amplicons. Always use pipette tips with filters to avoid contaminating PCR reagents and master mixes.
10. Prepare 1 to several mL of a 2× PCR master mix *without* DNA polymerase if large numbers of isolates are being DNA typed. Of the few DNA polymerases we have tested, Platinum® *Taq* DNA Polymerase High Fidelity (Invitrogen) consistently gave the most uniform amplicons, and it is supplied with 10× buffer and 25 mM $MgSO_4$. Even though the 2× PCR master mix for each locus only differs by the forward and reverse PCR primers,

the volume of each ingredient should be written in a lab notebook as it is added to avoid errors. If a pipetting error is suspected once the master mix had been prepared, it should weigh 1 g after a top-loading balance is tared to zero with an empty microfuge tube. It is good practice to include a no DNA negative control in each PCR experiment to make sure the PCR master mix is not contaminated with DNA.

11. Purchase a PCR grade (Na salt) deoxynucleoside triphosphate set (100 mM dATP, dCTP, dGTP, dTTP), and if needed, prepare several mL of the dNTP mixes by adding 100 μL of each to 1.5 mL microfuge tubes containing 600 μL sterile molecular grade H_2O. Vortex to mix and store at −20 °C.

12. Portions of *TEF1*, *RPB1*, and/or *RPB2* are PCR amplified and sequenced to DNA type *Fusarium* isolates (Table 1; Fig. 1). If there are time and budgetary constraints, a partial *TEF1* sequence obtained with the reverse internal sequencing primer EF22i can be used to sequence through the 3 large phylogenetically informative introns in the 5′ half of the amplicon (Table 1; Fig. 1). When large numbers of strains are being typed, PCR master mixes for each locus are dispensed from a V-shaped trough using an 8-channel micropipette. After use, the plastic troughs are rinsed with sterile Type II H_2O and dried promptly so that they can be reused.

13. Of the numerous recipes for preparing gel loading dye, we selected bromophenol blue (BPB) in glycerol because this dye migrates at about 500 bp in a 1.5% agarose gel run in 1× TAE buffer. This is slightly faster than the shortest amplicon, *TEF1*, which is ~650–700 bp.

14. EtBr is a potent mutagen and should be handled with extreme care by wearing a fastened lab coat, gloves, and safety goggles (i.e., PPE). Because of its high toxicity we never incorporate it in agarose gels, although this is done in some labs to save the time required to stain the gel after it has been run. We recommend preparing 100 mL of a 10 mg/mL stock solution in a dark-brown screw-capped bottle if one is available and be aware that the solution will need to be stirred for several h on a magnetic stir plate before it is completely dissolved. To minimize exposure, leave the stir bar in the bottle. The stock solution should be stored in a secondary container such as a beaker at room temperature or in 4 °C refrigerator and only handled wearing PPE. For a safer alternative, we recommend using SYBER Green I to stain gels.

15. As a faster alternative to ExoSAP-IT, ExoSAP-IT Express (Affymetrix Inc.; Santa Clara, CA, USA), or Exo-CIP Rapid PCR Cleanup Kit (New England BioLabs, Ipswich, MA, USA), both degrade dNTPs and PCR primers in 4 min at

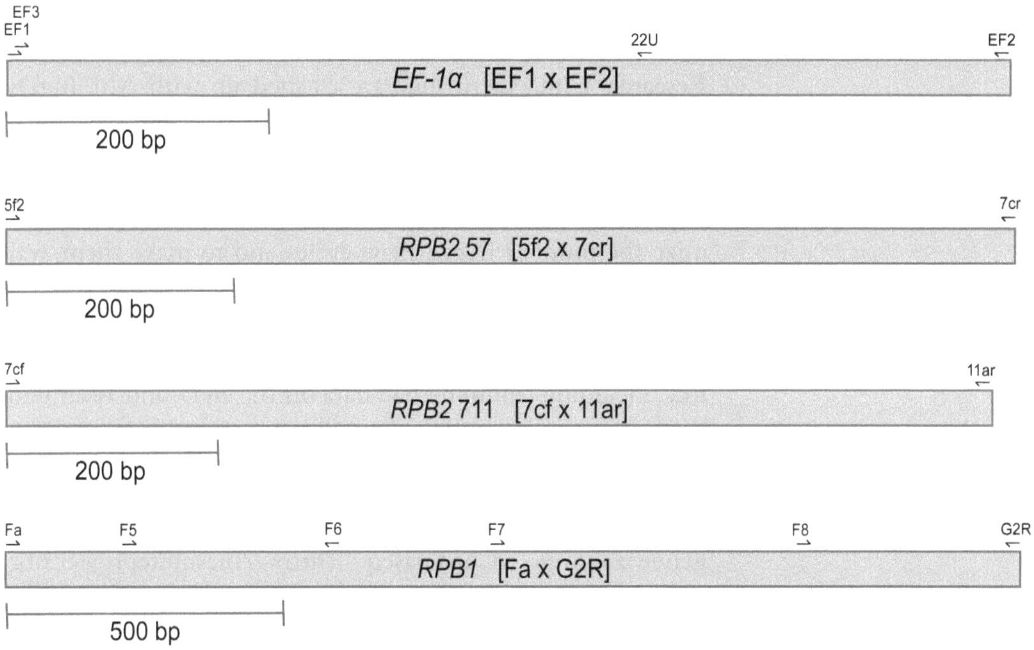

Fig. 1 Half arrows identify location and orientation of PCR and sequencing primers for translation elongation factor 1-α (*TEF1*) and DNA-directed RNA polymerase II largest (*RPB1*) and second-largest subunit (*RPB2*)

37 °C, using a thermostable calf intestinal phosphate and exonuclease 1, followed by inactivation at 80 °C for 1 min. For a less expensive alternative, use the exonuclease I and shrimp alkaline phosphatase cleanup protocol provided by New England BioLabs (https://international.neb.com/-/media/nebus/files/application-notes/enzymatic-pcr-cleanup-exo_sap-app-note.pdf?rev=57a419a3155043c89d7365eda33b2ae9). These reactions should be set up in the PCR product contaminated area of the lab. If large numbers of amplicons are being processed, the cost per sample can be reduced by purchasing reagents in bulk and experimentally determining the minimal volume needed to obtain desirable results. Protocols for both kits recommend adding 2 μL reagent to 5 μL amplicon; however, we have found that 2 μL of ExoSAP-IT is enough to purify 10 μL of amplicon. If these kits are purchased in bulk (i.e., 1 or 10 mL), aliquot in smaller volumes for storage at −20 °C to minimize the number of freeze-thaw cycles. Before cleaning amplicons for DNA sequencing, bear in mind that the protocol provided uses 2 μL/sequencing reaction. The three loci we routinely sequence for *Fusarium* (Table 1; Fig. 1) require 2 (*TEF1*) or 4 sequencing primers (*RPB1*, *RPB2*), or a minimal of 4 and 8 μL clean amplicon, respectively. Practice will dictate whether a larger volume of amplicon should be cleaned up to enable resequencing with one or more of the primers.

16. We sequence amplicons in our lab, using skirted 96-well plates, and then bring them to a core facility in our ARS-USDA Research Unit where they are cleaned up with ABI BigDye XTerminator before running on an ABI 3730 48-capillary automated DNA sequencer. When a 1000 reaction kit is purchased (ThermoFisher), the BigDye Terminator and the 5× sequencing buffer are aliquoted into smaller volumes to minimize the number freeze-thaw cycles and to make them available to multiple scientists in our research group.
17. Unless you are paying a sequencing center to provide you with DNA sequences that have been edited so that they are error-free, including trimming bad data off the ends, and assembling them as a single contig, you will need software for viewing, editing, and assembling the .ab1 chromatograms. A selection of programs that are free or commercially available and some selected features can be found at the following: https://www.genewiz.com/. Chromaseq (http://mesquiteproject.org/packages/chromaseq/) is a free program that can be used to organize chromatograms into projects where they can be edited and assembled into contigs for conducting BLASTn queries of web-accessible databases. Two of the most popular commercially available programs include Sequencher (http://www.genecodes.com/) and Geneious (https://www.geneious.com/). However, prior to acquiring software for viewing, editing, and assembling .abi trace chromatograms, check with your DNA sequencing center to see if a sequence editor is available there for end users.
18. *Fusarium*-specific DNA sequence databases are available and include FUSARIUM-ID (http://isolate.fusariumdb.org/blast.php) and *Fusarium MLST* (http://fusarium.mycobank.org/), although more general databases could also be used such as NCBI GenBank (https://www.ncbi.nlm.nih.gov/nucleotide/).
19. Culturing strains in broth on a shaker compared with growth on agar requires less time th (ordinarily 2–4 days compared with 1 weeks or more) and enables easier and faster harvesting of cultures (esp. important for synchronizing timing when many isolates are being processed) and monitoring them for bacterial contamination. Regarding the latter, ordinarily if yeast-malt broth cultures are contaminated with bacteria, they will become clouded and outgrow *Fusarium* within 24 h. However, it is good practice to examine 10 μL of each culture for contamination with a compound microscope using a 40× or higher objective before it is harvested because growth of some fastidious bacteria does not cloud the medium even after 1 week. Note that most *Fusarium* species produce abundant microconidia that cloud the medium. Because we frequently

grow 50–100 cultures at the same time, growing them on a rotary shaker with racks that hold 50 mL disposable polypropylene disposable tubes enable high-throughput processing for DNA preps. Alternatively, glass test tubes or 125 mL sterile Erlenmeyer flasks may be used.

20. If large numbers of strains are being processed, we recommend using two separate Büchner funnels for harvesting mycelium on Whatman filter paper discs. Mycelium that collects beyond the margin of the disc should be washed onto the disc with a stream of Type II H_2O from a squeeze bottle with a nozzle. When the mycelium is still slightly moist, remove the disc and scrape it off with the tapered end of a stainless-steel spatula. If the genomic DNA preps are for PCR of marker loci, and not whole-genome sequencing, they can be collected in 1.5 mL microfuge tubes before attaching a Kimwipe square with a rubber band to cover the top.
21. Samples should be frozen for at least 1 h before placing on a freeze dryer to prevent free water from being sucked into the pump and contaminating the oil. After each run, the freeze dryer should be shut down so that the chamber can be cleaned, and to remove water from the line.
22. We dispense chloroform in a 50 mL screw-capped tube in a chemical fume hood so that it can be dispensed easily with a 100–1000 μL micropipette. Once the chloroform has been added to each DNA prep, and an emulsion has been obtained by vigorously shaking briefly, the sample can be spun in a microfuge on the lab bench.
23. Once 100 μL of sterile molecular grade H_2O is added to each genomic DNA prep, placing them in a heat block set at 60 °C for 30–60 min aids in dissolving most of the pellets. If some of the pellets haven't dissolved, drawing them up in a 20–200 μL micropipette with a filter pipette tip is an efficient way to break them up. Total genomic DNA preps should be stored in a −20 °C freezer when not in use. Many labs resuspend DNA preps in TE buffer (10 mM Tris-Cl pH 8.0, 1 mM EDTA pH 8.0), because EDTA chelates cations that nonspecific nucleases require to degrade DNA. However, TE buffer is not required if lab personnel possess sufficient laboratory skills and DNA is not left at room temperature longer than necessary.
24. For preparing the agarose gel, if a gel caster is used to clamp in the gel tray, use a leveling bubble to make sure the tray is level. Also check each comb when placed in the gel tray to make sure they are level. Determine in advance what volume of agarose is needed to ensure the teeth on the gel comb are covered by ~3 mm of 1.5% agarose. This can be done by determining what

volume of water is needed to cover the comb teeth. When melting 1.5% agarose in a microwave, make sure the cap is very loose, and monitor continuously to make sure that it doesn't boil over. Once the molten agarose has cooled to ~70 °C, pour the desired volume into a graduated cylinder, and then pour directly into the gel tray with combs in place. If air bubbles collect on the agar surface, use a pipette tip to move them to the edge of the gel. Once the gel has completely solidified, remove the combs by gently pulling them straight up. Using a timer to determine that the gel has completely solidified can avoid damaging it when the combs are removed. We increased the agarose from 1% to 1.5%, which significantly increased the ease of handling. Just covering gels with 1× TAE buffer increases the ease of loading them. A 12-channel 2–20 μL micropipette that never leaves the PCR contaminated area is used to dispense 10 μL of the 6× gel loading buffer to a recycled 96-well plate to which 2 μL of each amplicon is added. Use a V-shaped trough if large numbers of amplicons are being processed. Prior to loading samples, gently tap the plate on the laboratory bench to collect the amplicons at the bottom of the tubes and to dislodge any air bubbles. A 12-channel 0.5–10 μL micropipette is used to load 10 μL into every other well, leaving one well on either end for loading 6 μL of a suitable molecular weight marker such as the EZ load 100 bp ladder (Bio-Rad, Hercules, CA), and then visualized over a UV trans-illuminator (Fotodyne Inc.; Hartland, WI) after staining with ethidium bromide. The marker should contain fragments that are larger and smaller than the amplicons so that they can be sized. The molecular weight marker also serves as a positive control for staining the gel with EtBr. The same pipette tips can be used to load the next set of 12 reactions if they are rinsed in the 1× TAE running buffer. Running the gel with the power supply connected to a timer ensures sample run time is precisely controlled during every run.

25. Mark off an area in the lab dedicated for staining agarose gels with EtBr. We recommend using a plastic gel staining tank with an attached lid that is large enough to hold the largest gel electrophoresed. For every 100 mL distilled water in the tank, add one drop of the EtBr stock with a 9″ Pasteur pipette, and then flush the tip in the solution to remove most of the concentrated stock. Use a 9″ Pasteur pipette for this task rather than a micropipette to avoid contaminating the latter. The tank should be gently rocked back and forth a few times to obtain a uniform solution of EtBr. Avoid adding the EtBr stock to a tank containing a gel, and if this is done make sure it is added as far away from the gel as possible and mix the solution immediately so that the concentrated stock does not come in contact with and over stain the gel.

26. After the gel is strained for 15–20 min, wearing PPE use a flat angled spatula to gently lift one side of the gel out of the stain to avoid immersing your gloves in the EtBr. Transfer the gel to a plastic tray dedicated for destaining the gel in distilled water for 15 min, discard the water down a sink drain, carefully place the gel on a UV transilluminator, and make sure no bubbles are underneath the gel. Remove one glove and use this hand to manipulate the camera so that it is not contaminated with EtBr. Leave the gel on the UV transilluminator until an image of the gel accurately documents the results for a lab notebook or digital library. If the marker is poorly stained, when viewed and photographed over a UV transilluminator, add 1 drop of the EtBr stock/100 mL water in the straining tank, and restain the gel for another 15 min. The gel and all associated EtBr contaminated materials should be collected in a bin near the gel staining station dedicated for the accumulation and disposal of EtBr contaminated waste following approved guidelines. Once a gel image has been recorded and the gel has been placed in the waste bin, the transilluminator should be carefully wiped to remove all residual EtBr staining solution.
27. Because we routinely sequence one or more 96-well plates of amplicons, once the sequencing reaction mixtures are prepared in a 1.5 mL microfuge tube, vortexed to mix and spun down briefly, the mixture is transferred to a disposable V-shaped trough so that 8 μL can be dispensed into each well using a 12-channel micropipette. The troughs are rinsed with sterile Type II H_2O so that they can be reused. The plates are then moved from the clean PCR/sequencing set up area to the contaminated area where 2 μL of each purified amplicon is added to each well. Use a 12-channel micropipette if large numbers of amplicons are being processed. Sequencing reaction mixtures are collected at the bottom of each tube by spinning them briefly in a microplate centrifuge. Consult the thermocycler manual to determine whether a rubber mat or adhesive film should be used as a plate covering for PCR and cycle sequencing.
28. The BigDye XTerminator Purification Kit removes unincorporated BigDye terminators. SAM solution enhances performance of the XTerminator solution and stabilizes the post-purification reactions. Note it is important to mix the XTerminator solution until it is homogenous before using. Other tips include the need to use wide-bore pipette tips to thoroughly mix the reactions because the BDX beads quickly settle out of solution, and to carefully check to make sure all of the 96-wells in the plate are sealed by the MicroAmp clear adhesive film. Plates should be disposed of in hazardous waste containers when the sequencing run is completed.

29. Because DNA sequence-based identifications require a high level of confidence for all base calls, every effort should be made to ensure the .abi chromatograms are error-free before using a text file of the consensus sequences to conduct BLASTn queries of a web-accessible DNA sequence data base. Truly ambiguous base calls that cannot be corroborated by an opposite strand sequence should be called as "N" or deleted if they are near the end of the sequence. Correcting .abi chromatograms is an iterative process in that they should be checked when first imported into a program such as Sequencher by assembling sequences from the same isolate, then assembling corrected sequences from two closely related isolates, and then checking sequences after conducting BLASTn queries if gapped sequences are present in the subject and query sequence alignment. Once sequence chromatograms are imported in Sequencher, trim the ends to remove any Ns that cannot be corrected. Assemble sequences from the same isolate and resolve any differences. Then assemble contigs from different isolates using a "high" assembly parameter (e.g., 97–99% is a good starting point), and check every polymorphic site (typically the ends have the most errors). If the contigs from some isolates do not assemble, then lower the assembly parameters and assemble them with the contigs of one other isolate. Then check every polymorphic site. Dissolve contigs, and then assemble sequences from each isolate using a 100% assembly parameter. If necessary, rename the contig of each strain and create a new sequence from each consensus. Highlight all the consensus sequences, and try to assemble them using 100% sequence identity. Then export a sequence of each unique consensus as a fasta file for conducting BLASTn queries.

    Once .abi chromatograms are corrected, assembled into contigs and each unique consensus is exported as a fasta text file, conduct BLASTn queries of FUSARIUM-ID (http://isolate.fusariumdb.org/blast.php), *Fusarium MLST* (http://fusarium.mycobank.org/), and/or NCBI GenBank (https://www.ncbi.nlm.nih.gov/nucleotide/). We recommend (1) querying FUSARIUM-ID or *Fusarium MLST* before GenBank because many of the sequences in the latter are misidentified; (2) always recheck chromatograms for errors at gapped positions in the BLASTn alignments; (3) check species names associated with the top BLASTn hits to make sure they are up-to-date and applied consistently; (4) query FUSARIUM-ID and *Fusarium MLST* and collect DNA sequence data from two independent loci if a single locus identification is ambiguous; and, (5) where necessary, review published GCPSR studies to determine what % similarity indicates the query and subject might be conspecific. While no uniform % identity can be used to infer conspecificity, in *Fusarium* it generally falls between 99.5% and 100%. For this reason, query sequences and the top BLASTn hits need to be error-free.

Alternatively, multiple sequences can be exported from Sequencher 5.4.6 as a fasta-concatenated file. Once the file is exported, type Ctrl A to select the fasta files, copy and paste them in the GenBank "Enter Query Sequence" search window, and then click on the BLAST button. Once your batch BLASTn request has been processed, you will be given a page that has the results for the first sequence in your list. To access the results for the other sequences, use the drop-down list at the top of the screen. It is possible to retrieve the results of batch BLASTn queries by clicking on the "Download" link located on the upper part of the result page.

Acknowledgments

The authors thank Amy McGovern and Crystal Probyn for skilled technical assistance. The US Department of Agriculture Agricultural Research Service (USDA-ARS) National Program for Food Safety and National Science Foundation (NSF; DEB-1655980) provided funding for this study. This research was supported in part by an appointment to the ARS Research Participation Program administered by the Oak Ridge Institute for Science and Education (ORISE) through an interagency agreement between the US Department of Energy (DOE) and the US Department of Agriculture (USDA). ORISE is managed by ORAU under DOE contract number DE-SC0014664. All opinions expressed in this paper are the authors' and do not necessarily reflect the policies and views of USDA, DOE, or ORAU/ORISE.

Disclaimer: *The mention of company names or trade products does not imply that they are endorsed or recommended by the US Department of Agriculture over other companies or similar products not mentioned. USDA is an equal opportunity provider and employer.*

References

1. Wingfield MJ, De Beer W, Slippers B, Wingfield BD, Groenewald JZ, Lombard L et al (2011) One fungus, one name promotes progressive plant pathology. Mol Plant Pathol. https://doi.org/10.1111/J.1364-3703.2011.00768.X
2. Geiser DM, Al-Hatmi A, Aoki T, Arie T, Balmas V, Barnes I et al (2020) Phylogenomic analysis of a 55.1 kb 19-gene dataset resolves a monophyletic *Fusarium* that includes the *Fusarium solani* species complex. Phytopathology 111:1064–1079
3. Taylor JW, Jacobson DJ, Kroken S, Kasuga T, Geiser DM, Hibbett DS et al (2000) Phylogenetic species recognition and species concepts in fungi. Fungal Genet Biol 31:21–32
4. Gerlach W, Nirenberg HI (1982) The genus *Fusarium* – a pictorial atlas. Mitt Biol Bundesanst Land-Forstwirtsch 209:1–406

5. Nelson PE, Toussoun TA, Marasas WFO (1983) *Fusarium* species: an illustrated manual for identification. Pennsylvania State University Press, University Park, PA
6. Leslie JF, Summerell BA (2006) The *Fusarium* laboratory manual. Blackwell Publishing, Ames, IA
7. O'Donnell K, Kistler HC, Cigelnik E, Ploetz RC (1998) Multiple evolutionary origins of the fungus causing Panama disease of banana: concordance evidence from nuclear and mitochondrial gene genealogies. Proc Natl Acad Sci U S A 95:2044–2049
8. O'Donnell K, Sutton DA, Rinaldi MG, Sarver BAJ, Balajee SA, Schroers H-J et al (2010) Internet-accessible DNA sequence database for identifying fusaria from human and animal infections. J Clin Microbiol 48:3708–3718
9. O'Donnell K, Rooney AP, Proctor RH, Brown DW, McCormick SP, Ward TJ et al (2013) *RPB1* and *RPB2* phylogeny support an early cretaceous origin and a strongly supported clade comprising all agriculturally and medically important fusaria. Fungal Genet Biol 52:20–31
10. Geiser DM, del Mar Jiménez-Gasco M, Kang S, Makalowska I, Veeraraghavan N, Ward TJ et al (2004) FUSARIUM-ID v 10: a DNA sequence database for identifying *Fusarium*. Eur J Plant Pathol 110:473–479
11. O'Donnell K, Ward TJ, Robert VARG, Crous PW, Geiser DM, Kang S (2015) DNA sequence-based identification of *Fusarium*: current status and future directions. Phytoparasitica 43:583–595
12. van Dam P, de Sain M, ter Horst A, van der Gragt M, Rep M (2018) Use of comparative genomics-based markers for discrimination of host specificity in *Fusarium oxysporum*. Appl Environ Microbiol 84:e01868–e01817

Chapter 2

Isolation of High-Molecular-Weight (HMW) DNA from *Fusarium oxysporum* for Long-Read Sequencing

Zunyong Liu, Eun-Gyu No, Catherine Danmaigona Clement, Ping He, and Libo Shan

Abstract

In many instances *Fusarium oxysporum* genomes are complex and challenging to assemble mainly due to the increased number of repetitive elements and variable numbers of supernumerary chromosomes, which are primarily associated with pathogenicity. Hence, to obtain the accurate *F. oxysporum* genome assembly and high-resolution sequence information, protocols for versatile, reliable, and high recovery of high-quality DNA for diverse sequencing platforms are instrumental. Here, we describe two protocols for the isolation of DNA from isolates of *F. oxysporum*. One is a quick and easy method for high-throughput extraction of DNA to rapidly screen diverse isolates by marker-assisted PCR analysis. Another is to harvest high-quality and high-molecular-weight DNA for whole-genome sequencing. In addition, we also include a library preparation protocol optimized for the third-generation sequencing technology using the portable MinION device to obtain long-read sequences. These protocols can be potentially further applied for all *Fusarium* spp. and other fungal pathogens, including *Verticillium*.

Key words *Fusarium oxysporum* f. sp. *vasinfectum* (*Fov*), Pathogenic races, DNA isolation, MinION sequencing

1 Introduction

MinION nanopore sequencing technology with the portable MinION device, released by Oxford Nanopore Technologies (ONT) in 2014, is a third generation-sequencing platform that can operate at higher speed and produce longer reads with a reduced cost [1–3]. MinION sequencing generates the DNA sequence data via monitoring the changes in electrical conductivity after the DNA strands pass through a biological pore [2, 4]. It has been shown that MinION is capable of generating the read lengths from several hundred bases to hundreds of thousands of bases and can span repetitive sequences and close gaps in short-read data to help assemble the genomic regions acquired from horizontal gene

Jeffrey Coleman (ed.), *Fusarium wilt: Methods and Protocols*,
Methods in Molecular Biology, vol. 2391, https://doi.org/10.1007/978-1-0716-1795-3_2,

transfer [1, 2]. Besides, the MinION device is small and portable equipment (about 90 grams) with a low price and can display data in real-time when connected with a computer. Thus, it is widely used for pathogen surveillance and disease diagnostics [1, 5]. The length of input DNA is a critical factor to increase the read length during MinION sequencing [4]. The *F. oxysporum* genome contains more repetitive sequence and transposable elements than that of its most closely related species *F. verticillioides* and *F. graminearum* [6]. Here, we describe a method to generate high-molecular-weight DNA from *F. oxysporum* for MinION sequencing (Fig. 1).

2 Materials and Equipment

2.1 Materials

1. *Fusarium oxysporum* isolates of interest for genome sequencing.
2. Potato Dextrose Agar.
3. YEPD: 20 g peptone, 20 g glucose, 10 g yeast extract in 1 L ddH_2O, autoclaved for 20 min.
4. Microcloth.
5. 100 mg/mL RNase solution: 0.1 g RNase in 1 mL sterile ddH_2O.
6. 50 mg/mL proteinase K: 50 mg proteinase in 1 mL sterile ddH_2O.
7. Blood & Cell Culture DNA Maxi Kit (10) (Qiagen) containing the QBT, QC, G2, and QF buffer.
8. 1% agarose: 1 g agarose in 100 mL 1× TBE solution.
9. 5× TBE buffer: 54 g Tris-base, 27.5 g boric acid, 20 mL 0.5 M EDTA (pH 8.0).
10. Ethidium bromide.
11. Liquid nitrogen.
12. Mortar and pestle.
13. Test tube rack.
14. DNA marker.
15. Isopropanol.
16. 75% ethanol.
17. Lambda DNA ladder marker.
18. Elution buffer (EB): 10 mM Tris–HCl pH 8.0.
19. TE buffer: 10 mM Tris–HCl pH 8.0, 1 mM EDTA.
20. 1.5 mL and 2 mL microcentrifuge tubes.
21. 1.5 mL DNA LoBind Microcentrifuge Tubes (Eppendorf).

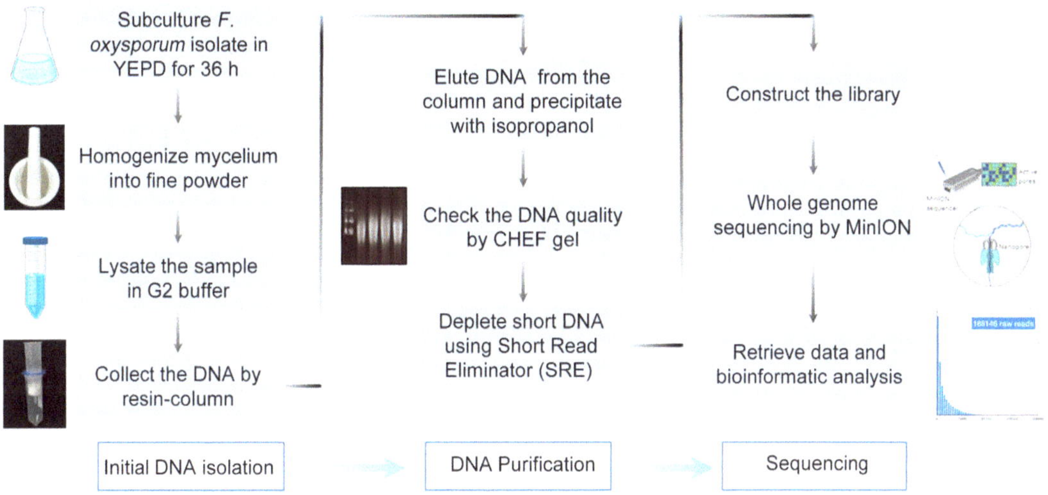

Fig. 1 The workflow of the main steps for the isolation and purification of high-quality *F. oxysporum* DNA and library preparation for MinION sequencing

22. 200 μL wide-bore pipette tips.
23. Qubit dsDNA BR Assay Kit.
24. Qubit dsDNA HS Assay kit.
25. Blunt/TA Ligase Master Mix (New England Biolabs).
26. NEBNext® Quick Ligation Reaction Buffer (New England Biolabs).
27. NEBNext® FFPE DNA Repair Mix (New England Biolabs).
28. NEBNext® Ultra II End Repair/dA-Tailing Module (New England Biolabs).
29. NEBNext® Quick Ligation Module (New England Biolabs).
30. Agencourt AMPure XP beads.
31. Magnet stand.
32. Ligation Sequencing Kit (Oxford Nanopore Technologies).
33. Flow cell for MinION (Oxford Nanopore Technologies).
34. Short Read Eliminator kit (Circulomics).

2.2 Equipment

1. Shaker.
2. Chamber.
3. Water bath.
4. Centrifuge.
5. MinION device (Oxford Nanopore Technologies).
6. Computer with specifications to run MinION device.
7. NanoDrop 2000 UV/Vis spectrophotometer (Thermo Fisher Scientific).

8. Vortexer.
9. Rotator mixer.
10. Thermal cycler.
11. Geno/Grinder tissue homogenizer (SPEX SamplePrep).
12. Qubit 3.0 fluorometer.
13. CHEF-DR®III Pulsed Field Electrophoresis system (Bio-Rad).

3 Methods

3.1 DNA Preparation

3.1.1 Prepare the DNA for PCR

1. Subculture the *F. oxysporum* strains from glycerol stocks on PDA plates with 100 μg/mL Kanamycin or 100 μg/mL Ampicillin to avoid contamination using a sterile toothpick.
2. Put the plates in a 25 °C chamber with a 12/12 dark-light period and culture for 3–5 days.
3. Collect fresh mycelium from the PDA plate in a 2 mL microcentrifuge tube.
4. Add 500 μL DNA lysis buffer and an iron ball with 4 mm diameter to the 2 mL microcentrifuge tube containing the mycelium, and grind for 2 min with the Geno/Grinder.
5. Incubate in a 50 °C water bath for 15 min.
6. Centrifuge at 20,000 × *g* for 5 min at 25 °C (*see* **Note 1**).
7. Transfer 300 μL of the supernatant to a new 1.5 mL tube, and add an equal volume of isopropanol (*see* **Note 2**).
8. Centrifuge at 20,000 × *g* for 5 min at 4 °C and discard the supernatant.
9. Wash the pellet with 75% ethanol with centrifuge for 2 min at 4 °C.
10. Dry the tube at 25 °C for about 30 min.
11. Add 100 μL ddH_2O to dissolve the DNA.

3.1.2 Prepare the DNA for MinION Sequencing

1. Pick up three PDA plugs (5 mm diameter) from the edge of the *F. oxysporum* isolate to be sequenced, and put them in YEPD medium, with gentle shaking at 180 rpm for 36 h.
2. Collect the mycelium with a micro cloth.
3. Wash the mycelium using ddH_2O and dry it with a paper towel, and quickly measure the half-dry mycelium weight.
4. Quickly freeze the mycelium in liquid nitrogen. The mycelium can be stored at −80 °C until use.
5. Pre-warm a water bath and the 15 mL G2 buffer in a 50 mL tube to 50 °C.

6. Grind the mycelium with liquid nitrogen into a fine powder using mortar and pestle.
7. Transfer approximately 1 g of the mycelium powder into 15 mL pre-warm G2 buffer with 15 μL 100 mg/mL RNase, briefly vortex, and place in a 50 °C water bath for 1.5 h.
8. Add 75 μL 20 mg/mL proteinase K and return to the water bath for another 1.5 h.
9. Pre-chill a centrifuge to 4 °C and equilibrate a Qiagen Genomic-tip 500/G with 10 mL QBT buffer by gravity flow (*see* **Note 3**).
10. Centrifuge the sample at 4 °C at 3200 × *g* for 10 min (*see* **Note 4**).
11. Promptly transfer the supernatant into the pre-equilibrated Qiagen Genomic-tip 500/G, and allow the supernatant to pass through the resin by gravity flow.
12. Wash the Qiagen Genomic-tip with 15 mL of QC buffer twice by gravity flow through.
13. Elute the DNA in the resin with 15 mL of pre-warmed to 50 °C QF buffer by gravity flow.
14. Add an equal volume of isopropanol to the eluted DNA to precipitate genomic DNA and gently invert the tube 10–20 times.
15. Remove the flocculent genomic DNA from the solution, and transfer it to a 1.5 mL microcentrifuge tube (*see* **Note 5**).
16. Centrifuge at 4 °C at 20,000 × *g* for 10 min.
17. Discard the supernatant and wash the pellet twice with 75% ethanol.
18. Dry the DNA at room temperature for 10–30 min.
19. Suspend the DNA in 500 μL TE buffer (*see* **Note 6**).

3.2 Check the DNA Quality by CHEF Gel

1. Measure the concentration of all DNA samples with a NanoDrop, and adjust the concentration to be consistent.
2. Prepare 0.5× TBE for 1% pulse-field gel and running buffer.
3. Load the genomic DNA and lambda DNA ladder marker on the gel.
4. Perform pulsed-field gel electrophoresis using pre-cooled 0.5× TBE at 5.4 V/cm with a 120–180 s pulse time gradient for 26 h at 12 °C.
5. Stain the gel with 1% ethidium bromide for 30 min, then distain in water for 30 min, and image resulting the gel with a suitable imaging system.

3.3 MinION Sequencing

3.3.1 Construction of MinION DNA Library

The purified DNA can be used to sequence *F. oxysporum* genomes with the Oxford Nanopore Technologies MinION instrument. Two to four genomes can be multiplexed by indexing with barcode adaptors, and the library is constructed with the Ligation Sequencing Kit (LSK109). The multiplexed library was run on a flow cell (R9.4.1). The manufacturer's protocol is followed with two modifications. One of the modifications was that a trace of short DNA was depleted using Short Read Eliminator (SRE) size selection technology. The other modification is that the sample at the cleanup steps in the protocol was diluted with EB elution buffer to prevent bead clumping during the AMPure XP purification. Ideally to recover about 1 μg of the final library DNA.

3.3.2 SRE Purification

1. Adjust the DNA sample to a total volume of 60 μL and a Qubit DNA concentration of 150 ng/μL. Pipette the sample into a 1.5 mL Eppendorf DNA LoBind tube.
2. Add 60 μL of SRE Buffer to sample 1. Mix thoroughly by gently tapping the tube.
3. Centrifuge at 18,000 × *g* for 30 min at room temperature.
4. Carefully remove the supernatant from the tube without disturbing the DNA pellet.
5. Add 300 μL of the 70% EtOH wash solution to the tube. Do not tap or mix after adding the 70% EtOH. Centrifuge at 18,000 × *g* for 2 min at room temperature.
6. Remove the wash solution from tube without disturbing the DNA pellet.
7. Repeat **steps 5** and **6**.
8. Add 50 μL of EB Buffer to the tube and incubate at room temperature for 20 min.
9. Analyze the recovery and purity of the DNA by NanoDrop and Qubit.

3.3.3 MinION Library Construction (See **Note 7**) and Sequencing

1. Transfer 2.5 μg of the SRE-purified DNA into a 1.5 mL DNA LoBind tube. Adjust the volume to 48 μL with nuclease-free water.
2. In a 0.2 mL thin-walled PCR tube, set up 60 μL of the following DNA repair and End-Prep reaction:

   50 μL DNA sample (2.5 μg)
   3.5 μL NEBNext FFPE DNA Repair Buffer
   3.5 μL Ultra II End-prep reaction buffer
   3 μL Ultra II End-prep enzyme mix
   2 μL NEBNext FFPE DNA Repair Mix
   Set up multiple reactions simultaneously for multiplexing.

3. Mix well by pipetting using wide-bore pipette tips. Incubate at 20 °C for 5 min followed by 65 °C for 5 min using a thermal cycler,
4. Transfer the DNA sample to a clean 1.5 mL DNA LoBind tube.
5. Add 60 μL EB Buffer to the 60 μL of the DNA-repaired and end-prepped sample from **step 4**, and add 120 μL of AMPure XP bead solution (vortex AMPure XP beads before adding).
6. Incubate on a rotator mixer for 5 min at room temperature.
7. Collect the sample with a quick centrifugation and pellet on a magnet stand until eluate is clear. With the tube on the magnet stand, carefully remove the supernatant with a pipette.
8. Keep the tube on the magnet and wash the beads with 300 μL of freshly prepared 70% ethanol without disturbing the pellet. Remove and discard the ethanol using a pipette.
9. Repeat **step 8**.
10. Spin down and place the tube back on the magnet, pipette off any residual ethanol. Allow to dry for ~30 s, but do not over-dry the pellet.
11. Remove the tube from the magnetic stand and suspend the pellet in 25 μL nuclease-free water. Spin down and incubate for 2 min at room temperature.
12. Pellet the beads on magnet until the eluate is clear.
13. Remove and retain the 25 μL of eluate into a clean 1.5 mL DNA LoBind tube. Quantify 1 μL of the eluted DNA using the Qubit dsDNA HS Assay kit.
14. In a 1.5 mL DNA LoBind tube, set up 50 μL of the following DNA barcode ligation reaction:C

    22.5 μL DNA-repaired and end-prepped DNA

    2.5 μL Native Barcode

    25 μL Blunt/TA Ligase Master Mix

    Set up multiple reactions simultaneously for multiplexing.
15. Mix well by pipetting using wide-bore pipette tips. Incubate the reaction for 10 min at room temperature.
16. Add 50 μL EB buffer to the 50 μL of the barcode-ligated sample in **step 15**, and add 100 μL AMPure XP bead solution (vortex AMPure XP beads before adding).
17. Incubate on a rotator mixer for 5 min at room temperature.
18. Spin down the sample and pellet on a magnet stand. Keep the tube on the magnet, and pipette off the supernatant.
19. Keep the tube on the magnet and wash the beads with 300 μL of freshly prepared 70% ethanol without disturbing the pellet. Remove and discard the ethanol using a pipette.

20. Repeat **step 19**.
21. Spin down and place the tube back on the magnet and pipette off any residual ethanol. Allow to dry for ~30 s, but do not over-dry the pellet.
22. Remove the tube from the magnetic and suspend the pellet in 26 μL of nuclease-free water. Incubate for 2 min at room temperature.
23. Pellet the beads on a magnet until the eluate is clear.
24. Remove and retain 26 μL of the eluate containing the DNA library into a clean 1.5 mL DNA LoBind tube.
25. Quantify 1 μL of an eluted sample using the Qubit dsDNA HS Assay kit.
26. Thaw the Elution Buffer (10 mM Tris–HCl, pH 8.0), NEB Next Quick Ligation Reaction Buffer (5×), and Long Fragment Buffer (LFB) at room temperature, mix by vortexing, spin down and place on ice.
27. Spin down the T4 Ligase and the Adapter Mix II (AMII), and place on ice.
28. Using the pooled and barcoded DNA, set up 100 μL of adapter ligation reaction as follows:
    65 μL 2 μg pooled barcoded sample DNA
    5 μL Adapter Mix II (AMII)
    20 μL NEBNext Quick Ligation Reaction Buffer (5×)
    10 μL Quick T4 DNA Ligase
29. Mix gently by flicking the tube, and spin down. Incubate the reaction for 10 min at room temperature.
30. Add 100 μL of 1× NEB Next Quick Ligation Reaction Buffer (prepared by mixing 20 μL of 5× Quick Ligation Reaction buffer and 80 μL of nuclease-free water) into the 100 μL of the adaptor-ligated sample from **step 29**.
31. Add 100 μL of the AMPure XP bead solution (vortex AMPure XP beads before adding).
32. Incubate on a rotator mixer for 5 min at room temperature.
33. Place on the magnetic stand, allow beads to pellet, and pipette off supernatant.
34. Wash the beads by adding 250 μL of the Long Fragment Buffer (LFB). Flick the beads to resuspend, spin down, and then return the tube to the magnetic stand for pelleting. Remove and discard the supernatant using a pipette.
35. Repeat **step 34**.
36. Spin down and place the tube back on the magnet. Pipette off any residual supernatant. Allow to dry for 30 s, but do not over-dry the pellet.

37. Remove the tube from the magnetic stand and re-suspend the pellet in 32 μL Elution Buffer (EB). Spin down and incubate for 10 min at room temperature.
38. Pellet the beads on a magnet until the eluate is clear.
39. Pipette 31 μL of eluate containing the DNA library into a clean 1.5 mL DNA LoBind tube.
40. Quantify 1 μL of the barcoded and adapter-ligated DNA using the Qubit dsDNA HS Assay kit.
41. Dilute 430 ng of the adapter-ligated and barcoded library DNA in 12 μL nuclease-free water and load into the flow cell installed in MinION sequencer as the manufacturer's instruction (*see* **Notes 8** and **9**).

4 Notes

1. The DNA lysis buffer contains SDS. Hence, this step cannot be conducted at 4 °C.
2. Avoid pipetting the debris into a new tube. If there is still some debris in the supernatant, you can extend the centrifuge time and pipette less supernatant to a new microcentrifuge tube.
3. Flow begins automatically by reducing surface tension due to the presence of detergent (0.15% Triton® X-100) in the equilibration buffer. Allow the QIAGEN Genomic-tip to drain completely. Do not force out the remaining buffer, as this will necessitate restarting the flow with a syringe and adapter.
4. Do not apply the sample with mycelium to the genomic tip, as it will clog the column making it difficult for the sample to pass through the resin.
5. If there is not an adequate quantity of flocculent DNA, genomic DNA can be obtained by centrifugation and collecting the DNA pellet.
6. The DNA may be difficult to dissolve, and the sample can be left at 4 °C overnight. The DNA concentration should be about 300–800 ng/μL, with an $OD_{260/280}$ value about 1.80–1.85. This quantity is adequate for MinION sequencing. The DNA size is usually over 50 kb.
7. This protocol can produce enough DNA library to provide three library loads from a single preparation. A 1.2 μg DNA library was recovered, which can be used to load the flow cell up to three times with 50 fmole of the DNA library an estimated size of 12–13 kb on average. The most updated MinION sequencing protocol added a DNaseI cleaning of the flow cell pores to remove DNA clogging the pores before adding a new DNA library. This protocol is sufficient to produce enough

DNA library to follow the current sequencing run protocol. Using only1 μg of input DNA may generate an amount only for a single load. When the input DNA was increased to 2.5 μg, the magnetic beads may clump. This method has optimized the pelleting by diluting the bead solution without altering the buffer stringency.

8. A total of 11 Gb of sequence data can be obtained, which is the equivalent to cover about 180× depth of the *Fusarium* genome. It is suggested to combine multiple different genomic DNA samples in a single flow cell by indexing with different barcodes, called multiplexing. Eleven Gb of data permits multiplexing with 90×, 60×, and 45× genome coverage per barcoded sample with a 2-plex, 3-plex, or 4-plex set up, respectively, with the current protocol.
9. This protocol generated 21 kb in sequence length N_{50} and 260 kb in sequence for the largest read length. The size of the input DNA when prepped through the Genomic-tips ranges from 50 to 150 kb.

Acknowledgments

This work was partially supported by the USDA NIFA (2019-67028-29949, 2018-67013-28513) and Cotton Incorporated cooperative research agreement (18-311) to L.S.

References

1. Jain M, Fiddes IT, Miga KH, Olsen HE, Paten B, Akeson M (2015) Improved data analysis for the MinION nanopore sequencer. Nat Methods 12:351–356
2. Mikheyev AS, Tin MM (2014) A first look at the Oxford Nanopore MinION sequencer. Molecul Ecol Resour 14:1097–1102
3. Ashton PM, Nair S, Dallman T, Rubino S, Rabsch W, Mwaigwisya S, Wain J, O'grady J (2015) MinION nanopore sequencing identifies the position and structure of a bacterial antibiotic resistance island. Nat Biotechnol 33:296–300
4. Lu H, Giordano F, Ning Z (2016) Oxford Nanopore MinION sequencing and genome assembly. Genom Proteom Bioinf 14:265–279
5. Wang J, Moore NE, Deng Y-M, Eccles DA, Hall RJ (2015) MinION nanopore sequencing of an influenza genome. Front Microbiol 6:766
6. Ma L-J, Van Der Does HC, Borkovich KA, Coleman JJ, Daboussi M-J, Di Pietro A, Dufresne M, Freitag M, Grabherr M, Henrissat B (2010) Comparative genomics reveals mobile pathogenicity chromosomes in *Fusarium*. Nature 464:367–373

Chapter 3

Karyotyping of *Fusarium oxysporum* by Pulsed-Field Gel Electrophoresis and the Germ Tube Burst Method

Yu Ayukawa and Masatoki Taga

Abstract

In fungi, karyotyping is fundamental to understanding their genome organization. It is also essential to study various genome- or chromosome-related topics such as karyotype polymorphisms and supernumerary or pathogenicity chromosomes. Here, we describe the protocols of pulsed-field gel electrophoresis and the germ tube burst method for molecular and cytological karyotyping of *Fusarium oxysporum*. The combined use of the two methods is valuable for determining definitive and comprehensive karyotypes of these fungi.

Key words Karyotype, *Fusarium oxysporum*, Pulsed-field gel electrophoresis, Germ tube burst method, Chromosome, Genome

1 Introduction

A karyotype is the chromosomal complement of a nucleus in the eukaryotic cell, in which the number and features of chromosomes are discerned. Traditionally, the term karyotype has been used to indicate the cytological information obtained by light-microscopy (hereafter, called cytological karyotype). Today, the term is expanded to include the information of the complement of chromosomal DNA (chDNA) (hereafter, called molecular karyotype), where the chromosome number and size are replaced with the number of chDNAs and their physical sizes in the unit of base pairs. The information gathered from a karyotype, either cytological or molecular, is fundamental to analyzing the genome of eukaryotes and defining a taxonomic species. In this chapter, we describe protocols for pulsed-field gel electrophoresis (PFGE) and the germ tube burst method (GTBM) for molecular and cytological karyotyping of *Fusarium oxysporum*.

PFGE separates chDNAs on a gel, enabling the rapid determination of chromosome number, chromosome size, and genome

Jeffrey Coleman (ed.), *Fusarium wilt: Methods and Protocols*,
Methods in Molecular Biology, vol. 2391, https://doi.org/10.1007/978-1-0716-1795-3_3,

size, thereby, generating a molecular karyotype called electrophoretic karyotype (EK) [1]. While different types of apparatuses were used in the early stages [1, 2], the contour-clamped homogeneous electric field (CHEF) type is exclusively used presently. Numerous EKs have been published in fungi [2]; however, in light of critical evaluation a significant proportion of the previous EKs could be underestimates of the chromosome number and genome size due to a disregard of co-migration of similar-sized chromosomes and/or the upper resolution limit of large chromosomes [3]. Thus, decisive molecular karyotypes that are needed in genome projects are currently determined with optical mapping combined with genome sequencing (for *F. oxysporum* see [4]). Nevertheless, PFGE remains an essential research tool for genome- or chromosome-related issues such as karyotype polymorphism(s), construction of a chromosome-specific library, gene mapping to specific chromosomes, and detection of supernumerary (or B) and pathogenicity chromosomes in fungi.

GTBM is a cytological method developed in 1988 to observe mitotic metaphase chromosomes of fungi [5]. Before GTBM, fungal karyotypes had been determined by conventional light-microscopic cytology using a sample-squash technique (for *F. oxysporum* see [6]). But, such previous methods in cytology, particularly for analyzing fungal mitotic chromosomes, were likely to lead to incorrect results, hampered by the minute size of fungal chromosomes and persistence of the nuclear membrane during mitosis [7]. GTBM, on the other hand, utilizes the bursting of germ tube cells to discharge chromosomes outside of the cells and spread them on a glass slide surface (for the principle of GTBM, see Discussion in [8]). Consequently, chromosome specimens required for solid karyotyping, i.e., non-overlapping, enlarged mitotic chromosomes, can be prepared with the GTBM. In the 1990s, the original version of GTBM was improved by combining with fluorescence microscopy for clearer visualization of chromosomes and applying fluorescence in situ hybridization to the chromosomes [9].

It is noteworthy that PFGE and GTBM are complementary methods for karyotyping. As mentioned above, the major disadvantage of PFGE lies in that large chromosomes (>6 Mb) or similar-sized chromosomes are difficult to resolve with PFGE, resulting in the underestimation of the chromosome number (e.g., see [3]). However, such is not the case for GTBM. On the contrary, absolute size measurement of chromosomes is not possible with GTBM, whereas it is feasible with PFGE. To obtain definitive and comprehensive karyotypes of *F. oxysporum*, the combined use of PFGE and GTBM is recommended [10].

2 Materials

2.1 Protoplasting

1. NO_3 medium: 0.17% (w/v) yeast nitrogen base without amino acids, 3% (w/v) sucrose, 100 mM potassium nitrate.
2. Miracloth or Kimwipes.
3. Centrifuge.
4. Hemacytometer.
5. Potato dextrose broth (PDB): 200 g/L potato infusion, 2% (w/v) glucose.
6. 1.2 M $MgSO_4$.
7. Enzyme solution: 2% (w/v) Driselase (Sigma-Aldrich, MO, USA) and 2% (w/v) lysing enzyme from *Trichoderma harzianum* (Sigma-Aldrich) in 1.2 M $MgSO_4$.
8. 50 mL centrifuge tubes.
9. Sorbitol solution (pH 7.4): 1 M sorbitol, 50 mM $CaCl_2$, 10 mM Tris (pH 7.4).

2.2 Preparation of Agarose Plugs

1. STE: 1 M sorbitol, 25 mM Tris (pH 7.5), 50 mM EDTA.
2. SNT: 1 M sorbitol, 0.1 M NaCl, 10 mM Tris (pH 7.4).
3. Low melting agarose (Bio-Rad, CA, USA).
4. Disposable plug molds (Bio-Rad).
5. NDS: 0.5 M EDTA, 1% (w/v) N-lauroylsarcosine sodium salt, 10 mM Tris–HCl (pH 8.0).
6. Proteinase K.
7. 50 mM EDTA (pH 8.0).

2.3 Pulsed-Field Gel Electrophoresis (PFGE)

1. 0.5× TBE buffer (2.5 L): 44.5 mM Tris, 44.5 mM boric acid, 1 mM EDTA.
2. SeaKem Gold agarose (Lonza, Basel, Switzerland).
3. CHEF Mapper XA Pulsed Field Electrophoresis System (Bio-Rad).
4. CHEF DNA Size Markers (*Hansenula wingei* chromosomal DNA) (Bio-Rad).
5. 100 μg/mL ethidium bromide.
6. Distilled water.
7. UV transilluminator.

2.4 Southern Hybridization

1. Denaturing buffer: 3 M NaCl, 0.4 M NaOH (for Nytran SPC TurboBlotter kit).
2. Transfer buffer: 3 M NaCl, 8 mM NaOH (for Nytran SPC TurboBlotter kit).

3. Positively charged nylon membranes.
4. Nytran SPC TurboBlotter kit.
5. Filter paper (for Nytran SPC TurboBlotter kit).
6. 3MM Chr blotting paper (for Nytran SPC TurboBlotter kit).
7. 5× neutralizing buffer: 0.6 M Na_2HPO_4, 0.4 M $NaH_2PO_4 \cdot H_2O$.
8. UV crosslinker.
9. PCR DIG Probe Synthesis Kit.
10. Maleic acid buffer (pH 7.5): 0.1 M maleic acid, 0.15 M NaCl. Adjust pH with NaOH.
11. 10× blocking solution: 10% (w/v) blocking reagent (Roche) in the maleic acid buffer.
12. 20× SSC (pH 7.0): 3 M NaCl, 0.3 M sodium citrate.
13. Hybridization buffer: 5× SSC, 0.1% (w/v) N-lauroylsarcosine, 0.02% (w/v) sodium dodecyl sulfate (SDS), 1× blocking solution.
14. Low stringency wash buffer: 2× SSC, 0.1% (w/v) SDS.
15. High stringency wash buffer: 0.5× SSC, 0.1% (w/v) SDS (*see* **Note 1**).
16. Tween 20.
17. Anti-Digoxigenin-AP, Fab fragments.
18. Detection buffer (pH 9.5): 0.1 M Tris–HCl, 0.1 M NaCl.
19. CDP-Star.
20. Hybridization bags.
21. X-ray films.
22. Developer solution.
23. Acetic acid.
24. Fixer solution.

2.5 Germ Tube Burst Method

1. PDB: 200 g/L potato infusion, 2% (w/v) glucose.
2. Kimwipe.
3. Centrifuge.
4. Hemacytometer.
5. Pre-cleaned glass slides.
6. Diamond pen.
7. Paper cement in a tube.
8. 0.1 mg/mL poly-L-lysine hydrobromide in water.
9. 10 mg/mL thiabendazole (TBZ) stock solution in dimethyl sulfoxide. Store at −20 °C.

10. Coplin jars.
11. Methanol.
12. Acetic acid.
13. Vectashield mounting medium (Vector Laboratories, CA, USA).
14. 4′,6-diamidino-2-phenylindole dihydrochloride (DAPI): 100 μg/mL DAPI stock solution in distilled water. Store in a refrigerator. As working solution, prepare 1 μg/mL DAPI in Vectashield mounting medium and store at −20 °C until use.
15. Glass coverslips (18 × 32 mm).
16. Epifluorescence microscope equipped with ×40 and ×100 oil-immersion objectives and filter cubes for UV- and B-excitation.
17. 20× SSC (pH 7.4): 3 M NaCl, 0.3 M sodium citrate.
18. 100 μg/mL RNase A in 2× SSC.
19. Parafilm.
20. Ethanol.
21. DAPI/PI (propidium iodide): 1 μg/mL DAPI and 0.25 or 0.5 μg/ml PI in Vectashield mounting medium.
22. Nail enamel.
23. Image processing software (e.g., ImageJ or Adobe Photoshop).

3 Methods

3.1 PFGE

Steps 1–10 below should be carried out under sterile conditions.

3.1.1 Preparation of Protoplasts-Containing Agarose Plugs

1. Inoculate 100 mL of NO_3 medium in a 300 mL flask with *F. oxysporum* mycelial agar blocks cut from a culture grown on a PDA plate, and incubate on a reciprocal shaker (120 strokes/min) at 25 or 28 °C for several days to produce bud-cells.
2. Filtrate the culture through Miracloth or two-layers of Kimwipes into 50 mL conical tubes. Collect bud-cells by centrifugation (1500–1800 × *g* for 5–10 min).
3. Suspend bud-cells in sterile water and determine bud-cell concentration with a hemacytometer.
4. Add 4×10^8 bud-cells to 80 mL of PDB in a 300 mL flask, and incubate on a shaker (80 strokes/min) at 25 or 28 °C for 16 h (*see* **Note 2**).
5. Collect germinated cells by centrifugation (1300 × *g*, 10 min), wash the cells with 1.2 M $MgSO_4$, and store in 1.2 M $MgSO_4$ at 4 °C until its use for protoplasting.

6. Pellet the cells by centrifugation (1300 × *g*, 10 min) and suspend in the enzyme solution.
7. Incubate at 30 °C for 12–17 h (*see* **Note 3**) with gentle shaking to release protoplasts. Check the production of protoplasts under a microscope.
8. Filtrate the solution through Miracloth or six-layers of Kimwipes into a 50 mL centrifuge tube.
9. Mix the filtrate with four volumes of ice-cold sorbitol solution (*see* **Note 4**), and collect protoplasts as a pellet by centrifugation (1300–1800 × *g*, 10 min).
10. Suspend protoplasts in STE or SNT, and adjust the concentration to 4×10^8 protoplasts/mL.
11. Melt 1.2% (w/v) low melting agarose in STE, and maintain the gel at 50 °C on a heat block. The amount of the gel required depends on the volume of the protoplast suspension to which it will be mixed.
12. Mix protoplast suspension with an equal volume of the molten gel prepared above.
13. Pour the mixture into the disposable plug molds and solidify the gel plugs by placing it on ice.
14. Immerse the plugs in NDS containing proteinase K (2 mg/mL), and incubate at 50 °C with gentle shaking overnight.
15. Replace with fresh NDS containing proteinase K (2 mg/mL), and incubate another 24 h.
16. Wash the plugs with 50 mM EDTA (pH 8.0) three times for 2 h.
17. Store the plugs in 50 mM EDTA (pH 8.0) at 4 °C until use.

3.1.2 Separation of Chromosomal DNAs by PFGE

The procedures of PFGE described below have been used for *F. oxysporum* ff. spp. *lycopersici* and *conglutinans* and non-pathogenic *F. oxysporum* with the CHEF Mapper XA Pulsed Field Electrophoresis System to separate approximately 0.2–6.0 Mb chromosomes in a single run (Fig. 1a) [10]. When CHEF-DR II system is used, the readers may refer to Reference 11.

1. Prepare 150–200 mL of 1.0% SeaKem Gold agarose gel in 0.5× TBE buffer. Set aside 8–10 mL of the gel and keep molten at 50 °C to use in **step 3** below. Pour the remaining gel into the casting stand and remove the air bubbles.
2. Insert a comb into the gel to make the wells.
3. After the agarose solidifies, place the plugs containing the protoplasts and the CHEF DNA Size Markers into the wells, and fill in and seal the top of each well with the molten agarose set aside during **step 2**.

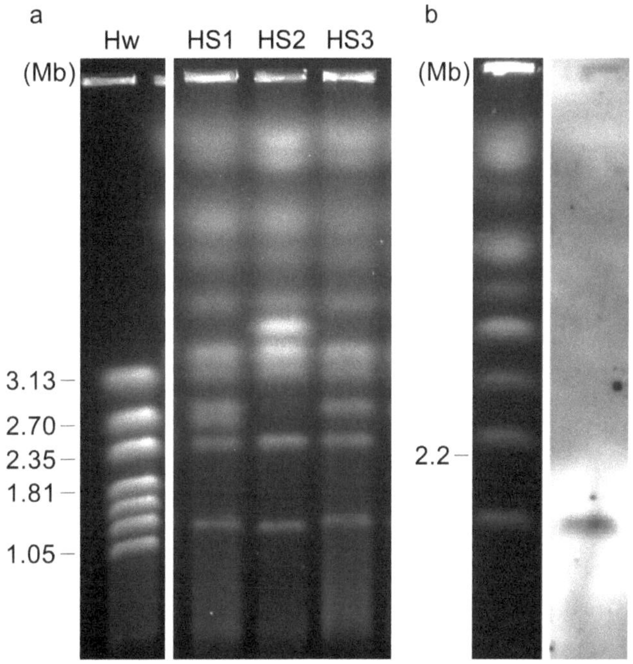

Fig. 1 Electrophoretic karyotype and PFGE-Southern hybridization of *Fusarium oxysporum*. (**a**) PFGE of *F. oxysporum* f. sp. *conglutinans*. HS1, HS2, and HS3, which are three mutants of strain Cong:1–1 whose dispensable chromosomes were artificially lost. Hw, *Hansenula wingei* chromosomal ladder. (**b**) PFGE-Southern hybridization of *F. oxysporum* f. sp. *lycopersici*. The virulence gene *SIX3* was located to a chromosomal band on the PFGE gel of strain IB-3

4. Place the gel attached to the platform in the electrophoresis chamber, and gently pour 0.5× TBE buffer into the chamber.
5. Perform electrophoresis at 7 °C with the following running conditions: 1.5 V/cm and 120° angle for 260 h with a switch time of 1200–4800 s (*see* **Note 5**).
6. Stain the gel with 100 μg/mL ethidium bromide for 30 min (*see* **Note 6**).
7. Destain the gel in distilled water for 30 min.
8. Visualize chromosomal DNA bands separated on the gel with a UV transilluminator.

3.1.3 Southern Hybridization

Several methods are available for blotting chromosomal DNA separated on a PFGE gel to a nylon membrane. We routinely use a Nytran SPC TurboBlotter kit because this kit requires less buffer and fewer filter papers.

1. Transfer chromosomal DNA to a positively charged nylon membrane with the Nytran SPC TurboBlotter kit.
2. After blotting, immerse the membrane in 1× neutralizing buffer for 5 min.

3. Bind DNA covalently to the membrane by UV cross-linking (120 mJ/cm^2).
4. Store the membrane in hybridization buffer at 4 °C until performing hybridization.
5. Prepare the DIG labeled-DNA probe with the PCR DIG Probe Synthesis Kit. Two μL of the labeled probe is necessary per mL of the hybridization buffer.
6. Incubate the membrane prepared at **step 4** in hybridization buffer for 2–3 h at a suitable temperature for hybridization (*see* **Note** 7).
7. Heat DIG labeled-DNA probe at 95 °C for 10 min and then quickly chill on ice.
8. Place the membrane in a hybridization bag, and add hybridization buffer containing the DIG-labeled probe to the bag. Heat-seal the bag and incubate at the desired temperature overnight.
9. Remove the hybridization buffer containing the DIG-labeled probe and replace with low stringency wash buffer, and allow to incubate for 15 min with gentle agitation at room temperature. Repeat the low stringency wash.
10. Incubate the membrane twice in preheated high stringency wash buffer for 15 min with gentle agitation at an appropriate temperature.
11. Wash the membrane with maleic acid buffer containing 0.3% (v/v) Tween 20 for 2 min.
12. Incubate the membrane in 1× blocking solution for 2–3 h with gentle agitation.
13. Incubate the membrane in 1× blocking solution supplemented with 0.2μL/mL Anti-Digoxigenin-AP, Fab fragments for 30 min.
14. Wash the membrane twice with maleic acid buffer containing 0.3% (v/v) Tween 20, each time for 30 min.
15. Incubate the membrane in the detection buffer for 3 min at room temperature with gentle agitation.
16. Put the membrane on a sheet of plastic wrap. Add 100-fold diluted CDP-Star to the membrane, cover the membrane with plastic wrap, and incubate in the dark for 10 min.
17. Transfer the membrane to a hybridization bag.
18. Expose the membrane to an X-ray film overnight.
19. Immerse the X-ray film in the developer solution.
20. Rinse the film in 3% (v/v) ascetic acid.
21. Immerse the film in fixer solution.
22. Take photographs of the film (Fig. 1b).

3.2 Cytological Karyotyping

The protocol below has been used with isolates of *F. oxysporum* ff. spp. *lycopersici* and *conglutinans* and non-pathogenic *F. oxysporum* [10]. For general protocols of GTBM applicable to other fungi, readers may refer to References 11, 12.

3.2.1 Preparation of Mitotic Chromosome Spreads by the Germ Tube Burst Method (GTBM)

1. Inoculate 100 mL of PDB in a 300 mL flask with two to three mycelial agar blocks cut from a PDA culture of *F. oxysporum*, and incubate on a reciprocal shaker (120 strokes/min) at 28 °C for 2 to 3 days to produce bud-cells.
2. Filtrate the liquid culture into 15 mL centrifuge tubes through two-layers of Kimwipe to remove mycelia, and collect bud-cells by centrifugation at 1800 × *g*.
3. Suspend bud-cells in PDB and adjust the concentration to 7.5×10^4 cells/mL with a hemacytometer.
4. Using a diamond pen, draw a rectangle (22 × 32 mm) on the central area of a glass slide. Put 20 μL of 0.1 mg/mL poly-L-lysine hydrobromide solution on the rectangle, and spread over the area with a micropipette (*see* **Note 8**). After drying the poly-L-lysine, squeeze paper cement out of a tube to trace the diamond-pen-drawn rectangle. The dried and solidified paper cement serves as an outer frame to keep the bud-cell suspension inside a rectangle area during slide incubation.
5. Place 150 μL of the bud-cell suspension in the rectangle area. Incubate the glass slide in a humidity chamber at 28 °C until the germ tubes originating from bud-cells begin to branch.
6. After incubation (usually for 12–13 h), check the proper growth of germ tubes from bud-cells under a microscope (*see* **Note 9**).
7. Discard the PDB inside the frame by gently removing with a micropipette, and replace it with 150 μL of fresh PDB containing 100 μg/mL of TBZ. During this process, the germinated bud-cells remain on the slide by adhering to the surface.
8. Resume incubation for another hour.
9. Peel off paper cement with forceps, and immerse the slide in distilled water in a Coplin jar for 2 min to wash off PDB.
10. Remove the slide from the jar and blot excess water from the sample area by touching a filter paper to its margin.
11. Slip the slides into the methanol–acetic acid mixture (9,1) in a proper container, and let it lie with the sample side up. Keep at room temperature for more than 20 min, during which time the germinated cells are burst and chemically fixed.
12. Remove the slide from the methanol–acetic acid, and wipe off excess methanol-acetic acid solution around the sample area with filter paper.

13. Pass the slide quickly through the flame of an alcohol lamp and burn off methanol–acetic acid.
14. Air-dry the slide in a fume hood, and store at room temperature until use.

3.2.2 Fluorescence Staining and Fluorescence Microscopy of Mitotic Chromosomes

DAPI Staining

1. Stain specimen with 15 μL of 1 μg/mL DAPI dissolved in Vectashield mounting medium, and cover with a glass coverslip.
2. Seal with nail enamel.
3. Using a ×40 or ×100 oil-immersion objective and UV-excitation with a filter cube for DAPI, observe the slides under a fluorescent microscope.
4. Select metaphase spreads that do not have chromosomes overlapping, and capture photographs with a ×100 objective and a CCD camera.

DAPI/PI Double Staining

1. Treat specimen with 70 μL of 100 μg/mL RNase A, and cover with a piece of parafilm (precut to the size of the cover glass) and incubate the slide at 37 °C for 1 h in a humidity chamber (*see* **Note 10**).
2. Rinse the slide in 2× SSC for 5 min.
3. Dehydrate the slide through a series of ethanol rinses (70%, 85%, and 95% (v/v)).
4. After air-drying of the slide, mount specimen with ca. 20 μL of the staining solution containing 1 μg/mL DAPI and 0.25–0.5 μg/mL PI in Vectashield mounting medium, and cover with a glass coverslip.
5. Seal with nail enamel.
6. Using a ×40 or ×100 oil-immersion objective and UV-excitation with a filter cube for DAPI, observe the slides under a fluorescent microscope, and select metaphase spreads without overlapping chromosomes.
7. Take photographs of the selected spreads under UV-excitation and B-excitation separately with a ×100 objective and a CCD camera (*see* **Note 11**).

3.2.3 Karyotyping

The basic karyotype including chromosome number and chromosome size (longitudinal axial length) can be determined using specimens stained with either DAPI or DAPI/PI. In the latter staining, heterochromatic regions are distinguishable by their intense, whitish fluorescence. The region of rDNA (nucleolar organizing region: NOR) can be identified as a protrusion from the apex of a particular chromosome, the so-called NOR chromosome. Although the karyotypic information can be obtained with the original picture(s) of the chromosome spread, making a chromosome alignment and

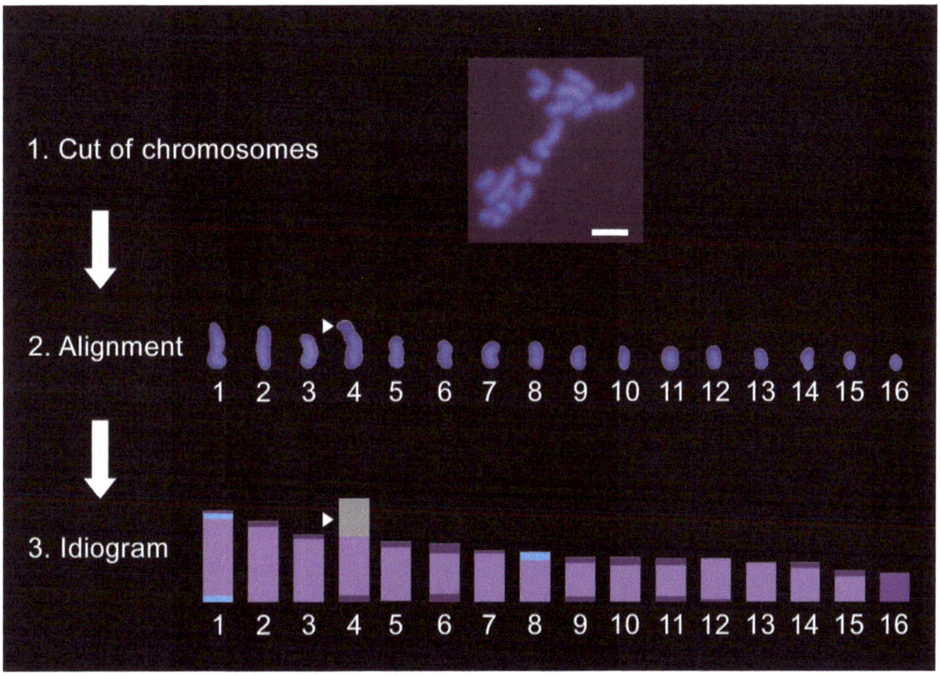

Fig. 2 Cytological karyotyping of *Fusarium oxysporum*. A metaphase plate stained with DAPI/PI is chosen for karyotyping and measured to determine the chromosomal size (longitudinal axial length) using ImageJ (upper panel). Individual chromosomes are isolated and aligned in order of size with Adobe Photoshop (middle panel). Draw an idiogram that represents the features of the individual chromosomes (lower panel). Scale bar in the metaphase plate indicates 2 μm. White arrowheads in the alignment and ideogram indicate the rDNA protrusion. In the ideogram, the purple and dark purple regions show affinity to both PI and DAPI, and the blue regions depict a strong affinity to DAPI

idiogram is preferable for sophisticated exhibition of the karyotype (Fig. 2).

1. Import the image of the metaphase chromosome spread into ImageJ, and measure the longitudinal axial length of each chromosome several times and calculate the average length (*see* **Note 12**).
2. Import the same image as mentioned above into Adobe Photoshop. Cut out each chromosome and align them in order of their longitudinal axial length.
3. Draw an idiogram based on the chromosome alignment.

4 Notes

1. If the GC-content of the probe is more than 50%, use 0.1× SSC containing 0.1% (w/v) SDS.

2. If it is necessary to produce a higher quantity of protoplasts, the culture can be scaled up. For instance, add 5×10^8 cells to 40 mL PDB in a 100 mL flask, and make replicates. Incubate the flasks on a shaker (175 rpm) at 25 °C for 12–17 h.
3. The incubation time necessary to generate enough protoplasts depends on the *F. oxysporum* isolates used. Some strains release plenty of protoplasts from germinated cells after 5 h of incubation.
4. As an alternative to the sorbitol solution, two volumes of 0.7 M NaCl can be used.
5. Change the buffer every 2 days.
6. Instead of ethidium bromide, non-toxic DNA dyes can be used for staining.
7. Calculate the hybridization temperature using the following equation:

   $T = \{16.6 \log [\text{Mol Na+}] + 0.41\ (\%\ G + C) + 81.5\} - 25\ °C$

   Mol Na + = Molar concentration of sodium in hybridization buffer

   T = optimal hybridization temperature

   % G + C = % of guanidine and cytidine residues in the probe sequence
8. Since commercially available poly-L-lysine solution adversely affects germination of the bud-cells of *F. oxysporum*, poly-L-lysine hydrobromide should be dissolved in water.
9. Overlapping growth of the long germ tubes should be avoided, in order to ensure that discharged chromosomes from different germ tubes do not mix.
10. PI binds to RNA as well as DNA. Since a large amount of RNA resides with the chromosomal specimens, the PI that binds to the RNA obstructs the observation of the chromosomes. To overcome this hindrance, RNase A is used to remove the RNA from the specimens.
11. Instead of B-excitation, G-excitation can be used.
12. When the NOR-chromosome is measured for its size, exclude the protrusion containing the ribosomal DNA (rDNA) region since the length varies depending on the specimen.

References

1. Mills D, McCluskey K (1990) Electrophoretic karyotypes of fungi: the new cytology. Mol Plant-Microbe Interact 3:351–357
2. Beadle J, Wright M, McNeely L, Bennett JW (2003) Electrophoretic karyotype analysis in fungi. Adv Appl Microbiol 53:243–270
3. Taga M, Murata M, Saito H (1998) Comparison of different karyotyping methods in filamentous ascomycetes–a case study of *Nectria haematococca*. Mycol Res 102:1355–1364
4. Ma LJ, Van Der Does HC, Borkovich KA et al (2010) Comparative genomics reveals mobile

pathogenicity chromosomes in *Fusarium*. Nature 464:367–373

5. Shirane N, Masuko M, Hayashi Y (1988) Nuclear behavior and division in germinating conidia of *Botrytis cinerea*. Phytopathology 78:1627–1630
6. Punithalingam E (1975) Cytology of some *Fusarium* species. Nova Hedwigia 26:275–303
7. Lu BCK (1996) Chromosomes, mitosis, and meiosis. In: Bos CJ (ed) Fungal genetics:principles and practice. Marcel Dekker, New York, pp 119–176
8. Taga M, Tanaka K, Kato S, Kubo Y (2015) Cytological analyses of the karyotypes and chromosomes of three *Colletotrichum* species, *C. orbiculare*, *C. graminicola* and *C. higginsianum*. Fungal Genet Biol 82:238–250
9. Taga M, Murata M (1994) Visualization of mitotic chromosomes in filamentous fungi by fluorescence staining and fluorescence in situ hybridization. Chromosoma 103:408–413
10. Ayukawa Y, Komatsu K, Taga M, Arie T (2018) Cytological karyotyping of *Fusarium oxysporum* by the germ tube burst method (GTBM). J Gen Plant Pathol 84:254–261
11. Mehrabi R, Taga M, Aghaee M, De Wit PJ, Kema GH (2012) Karyotyping methods for fungi. In: Plant fungal pathogens. Humana Press, New York, pp 591–602
12. Tsuchiya D, Taga M (2010) Fluorescence in situ hybridization for molecular cytogenetic analysis in filamentous fungi. In: Molecular and cell biology methods for fungi. Humana Press, New York, pp 235–257

Chapter 4

RNA-seq Data Analysis for Differential Expression

Navdeep Gill and Braham Dhillon

Abstract

Changes in the surrounding environment are mirrored by changes in the transcript profile of an organism. In the case of a plant pathogen, host colonization would be a challenge that triggers changes in transcript expression patterns. Determining the transcriptional profile could provide valuable clues on how an organism responds to defined stimuli, in this case, how a pathogen colonizes its host. Several robust data analysis methods and pipelines are available that can identify these differentially expressed transcripts. In this chapter we outline the steps and other caveats that are needed to run one such pipeline.

Key words RNA-seq, Transcriptome, Transcript profile, Data analysis, Pipeline, Differentially expressed genes, Splice-aware, HISAT2, StringTie, DESeq2

1 Introduction

The transcriptome represents the sum total of all the RNA in an organism. The first step in transcriptome sequencing or RNA-seq is extracting the RNAs from the cell and enrichment of target RNAs. Commonly, the protocol calls for either the enrichment of polyadenylated RNAs or depletion of ribosomal and transfer RNAs. The RNAs are then converted to complementary DNA (cDNA) and fragmented, followed by ligation of adapters, before these are sequenced [1].

The availability of transcript sequence information is crucial for supporting gene prediction and genome annotation efforts. Ab initio gene prediction programs use several informants to guide the prediction heuristics and accurately define gene features, like gene start site and exon-intron boundaries [2]. The accuracy of the prediction algorithms can be significantly increased by including transcript sequence information [3]. Moreover, RNA-seq datasets can be used to predict and reconstruct alternate transcripts and isoforms [4].

Jeffrey Coleman (ed.), *Fusarium wilt: Methods and Protocols*,
Methods in Molecular Biology, vol. 2391, https://doi.org/10.1007/978-1-0716-1795-3_4,

The population of total RNA changes as part of the normal growth process of an organism as well as in response to changes in its environment. Measuring these concerted changes in the transcript profile arising as a response to different stimuli can help understand the molecular mechanisms related to the fitness and survival of an organism. For a plant pathogen, like *Fusarium oxysporum* f. sp. *palmarum* that primarily attacks queen palms and Mexican fan palms in Florida [5], a change in the growing environment would lead to a shift in the expression of different groups of genes. This fungus can be isolated and successfully grown on an artificial medium containing simple sugars, like glucose. This constitutes a considerably different environment as opposed to the fungus growing on plant host tissues where complex carbohydrates need to be assimilated to obtain simple sugars. Thus, during the host-pathogen interaction, genes required for nutrient acquisition from complex plant carbohydrates as well as those required for suppression of the host immune response could be preferentially activated. The spatial and temporal resolution of this transcriptional response is key to deciphering the regulation of gene expression during host-pathogen interactions.

Quantifying the transcriptome has been greatly facilitated by the availability of next-generation sequencing (NGS) technologies that augment data generation and collection methods and at the same time accelerate the pace of discovery [6]. The current experimental bottleneck is not data generation but the speed and power of data analysis. The use of NGS methods has changed the landscape of computational resources and algorithms needed to assimilate and manage the volume of data generated in an experiment. Transcriptome data analysis methods, with sophisticated models and improved statistics, aimed at reducing false positive, and negatives are being constantly tested and developed. The NGS methods have revolutionized the "omics" field and profoundly expanded the scope of understanding long-standing biological questions.

The choice of tools and pipelines for RNA-seq data analysis depends on several factors, including the availability of a reference genome, sequencing platform, and read length. There have been numerous studies that compare and contrast different tools and methods on their ability to correctly identify differentially expressed genes with high confidence [7]. Even though the underlying statistics are robust, the power and suitability of these tools may need to be empirically determined for different datasets. The reference genome based HISAT2 [8] - StringTie [9] - DESeq2 [10] pipeline (Fig. 1) showcased here represents one of many possible combinations of different tools that can be used for RNA-seq data analysis.

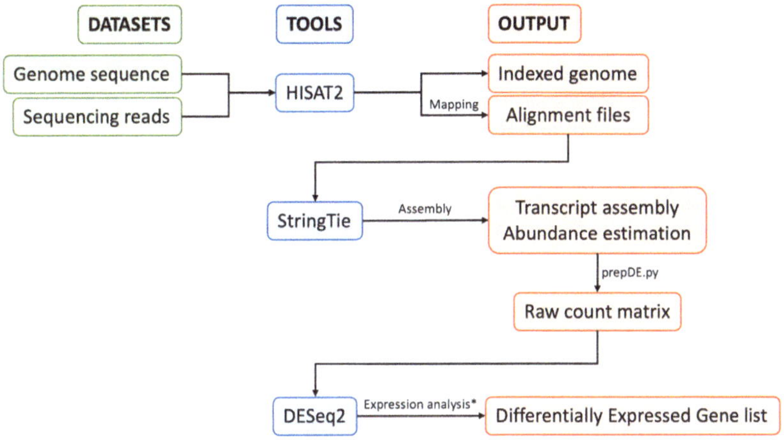

Fig. 1 An overview of the RNA-seq data analysis pipeline. This schematic shows the input datasets (green boxes), bioinformatics tools (blue boxes), and outputs (orange boxes) from each step in the pipeline that are needed for RNA-seq data analysis for differential gene expression (*see* **Note 9**)

2 Datasets and Directory Setup

1. The dataset for this experiment which contains a reference genome FASTA file, "genome.fa".
2. RNA-seq data FASTQ files (*see* **Note 1**) for six samples, i.e., two treatments, "cntrl" and "treat", with three replicates each, "01", "02", and "03". There are two files for each sample, suffix "_1" and "_2", that contain forward and reverse reads. Reads in the paired-end FASTQ files are arranged in the same order.
3. A master folder, "run01", with the following subfolders and files:

```
run01
 REF_genome
 genome.fa
 FASTQ_data
  cntrl01_1.fq, cntrl01_2.fq; cntrl02_1.fq, cntrl02_2.fq;
  cntrl03_1.fq, cntrl03_2.fq; treat01_1.fq, treat01_2.fq;
 treat02_1.fq, treat02_2.fq; treat03_1.fq, treat03_2.fq
 mapping
 assemble
 forDE
 mergeList.txt
 sampleAnno.csv
```

4. High performance computing with the following software/tools/packages (*see* **Note 2**):
   (a) HISAT2.
   (b) SAMtools.
   (c) StringTie.
   (d) DESeq2.
   (e) ggplot2.

3 Methods

3.1 RNA-seq Data Analysis

1. Index the reference genome (*see* **Note 3**). The reference genome "genome.fa" is indexed (*see* **Note 4**) using the following command:

```
hisat2-build -p 16 genome.fa genomes
```

2. Align reads to the reference. Paired-end datasets are aligned to the indexed reference genome, as below (*see* **Note 5**). The output files are housed in the "mapping" folder.

```
hisat2 -p 12 --dta -x genome -1 cntrl01_1.fastq -2 ctrl01_2.
fastq -S cntrl01_genome.sam
hisat2 -p 12 --dta -x genome -1 cntrl02_1.fastq -2 ctrl02_2.
fastq -S cntrl02_genome.sam
hisat2 -p 12 --dta -x genome -1 cntrl03_1.fastq -2 ctrl03_2.
fastq -S cntrl03_genome.sam

hisat2 -p 12 --dta -x genome -1 treat01_1.fastq -2 treat01_2.
fastq -S treat01_genome.sam
hisat2 -p 12 --dta -x genome -1 treat02_1.fastq -2 treat02_2.
fastq -S treat02_genome.sam
hisat2 -p 12 --dta -x genome -1 treat03_1.fastq -2 treat03_2.
fastq -S treat03_genome.sam
```

3. Convert the "SAM" to "BAM" and sort the files. SAMtools [11], the tool used for this operation, converts the large, human-readable Sequence Alignment Map (SAM) file format to a smaller, compressed Binary Alignment Map (BAM) format. SAM files can be deleted after this step to save space (*see* **Note 6**).

```
samtools sort -@ 12 -o cntrl01_genome.bam ctrl01_genome.sam
samtools sort -@ 12 -o cntrl02_genome.bam ctrl02_genome.sam
samtools sort -@ 12 -o cntrl03_genome.bam ctrl03_genome.sam
```

```
samtools sort -@ 12 -o treat01_genome.bam treat01_genome.sam
samtools sort -@ 12 -o treat02_genome.bam treat02_genome.sam
samtools sort -@ 12 -o treat03_genome.bam treat03_genome.sam
```

4. Assemble and merge the transcripts (*see* **Notes 4** and 7) for each sample with the command lines below. The output files are housed in the "`assemble`" folder.

```
stringtie cntrl01_genome.bam -l cntrl01 -p 12 -o cntrl01_ge-
nome.gtf
stringtie cntrl02_genome.bam -l cntrl02 -p 12 -o cntrl02_ge-
nome.gtf
stringtie cntrl03_genome.bam -l cntrl03 -p 12 -o cntrl03_ge-
nome.gtf

stringtie treat01_genome.bam -l treat01 -p 12 -o treat01_gen-
ome.gtf
stringtie treat02_genome.bam -l treat02 -p 12 -o treat02_gen-
ome.gtf
stringtie treat03_genome.bam -l treat03 -p 12 -o treat03_gen-
ome.gtf
```

5. For merging, the textfile "`mergelist.txt`", containing the names of the six "`gtf`" files created in the previous step, one file name per line,

```
stringtie --merge -p 12 -o stringtie_merged.gtf mergelist.txt
```

6. Estimate abundance. Two flags (*see* **Note 8**) "`-e`" that only estimate the abundance of given reference transcripts and "`-B`" that generates a Ballgown style sub-directory structure are needed for the next step.

```
stringtie -e -B -p 8 -G stringtie_merged.gtf -o forDE/cntrl01/
cntrl01_genome.gtf cntrl01_genome.bam
stringtie -e -B -p 8 -G stringtie_merged.gtf -o forDE/cntrl02/
cntrl02_genome.gtf cntrl02_genome.bam
stringtie -e -B -p 8 -G stringtie_merged.gtf -o forDE/cntrl03/
cntrl03_genome.gtf cntrl03_genome.bam

stringtie -e -B -p 8 -G stringtie_merged.gtf -o forDE/treat01/
treat01_genome.gtf treat01_genome.bam
stringtie -e -B -p 8 -G stringtie_merged.gtf -o forDE/treat02/
treat02_genome.gtf treat02_genome.bam
stringtie -e -B -p 8 -G stringtie_merged.gtf -o forDE/treat03/
treat03_genome.gtf treat03_genome.bam
```

7. Extract raw-counts matrix for transcripts

   A script, "prepDE.py," provided in the StringTie [9] package can be used to generate a count matrix for transcripts. As mentioned above, StringTie [9] needs to be used with two flags, "-e" and "-B," for this script to work. The matrix generated by this script can be imported into the R package, DESeq2 [10], for downstream analysis.

```
prepDE.py -i forDE
```

3.2 Differential Gene Expression

The R [12] Bioconductor [13] package, DESeq2 [10], is used to identify differentially expressed genes. A comma-separated two-column csv file, "sampleAnno.csv", that lists the sample name and annotation is needed.

```
Id Group
cntrl01 Control
cntrl02 Control
cntrl03 Control
treat01 Treatment
treat02 Treatment
treat03 Treatment
```

The sample list in "sampleAnno.csv" should be in the same order as the columns in the count matrix file.

1. Load the package.

```
library("DESeq2")
```

2. Load the count matrix file.

```
countData <- as.matrix(read.csv("gene_count_matrix.csv",row.
names="gene_id"))
colData <- read.csv("sampleAnno.csv", sep=",", row.names=1)
```

3. The columns in the count matrix and sample annotation file need to be in the same order.

```
all(rownames(colData) %in% colnames(countData))
countData <- countData[, rownames(colData)]
all(rownames(colData) == colnames(countData))
```

4. The count matrix is then loaded into DESeq2 as a DESeqDataSet (*see* **Note 9**).

```
dds <- DESeqDataSetFromMatrix(countData, colData, ~Group)
```

5. Next, the differential expression analysis is carried out and result tables are generated (*see* **Note 10**).

```
dds <- DESeq(dds)
res <- results(dds)
```

6. The results are ordered by p-adjusted values (*see* **Note 11**).

```
(resOrdered <- res[order(res$padj), ])
```

7. The next commands generate an unfiltered output with additional columns containing normalized counts for all genes (*see* **Note 12**), along with the regular output from the DESeq2 package. These results can then be filtered on the basis of $\log_2$ fold change and p-adj values (*see* **Note 13**).

```
resdata <- merge(as.data.frame(res), as.data.frame(counts
(dds,normalized=TRUE)), by="row.names", sort=FALSE)
names(resdata)[1] <- "Gene"
write.csv(resdata, file="diffexpr-results.csv")
```

8. Additionally, a principal component analysis (PCA) can help visualize the overall variation in the samples. Prior to PCA, the data needs to be transformed to reduce the dependence of the variance on the mean. In the case below, a "variance stabilizing transformation" (vst) is used.

```
vsd <- vst(dds, blind = FALSE)
pcaData <- plotPCA(vsd, intgroup=c("Group"), returnData=TRUE)
percentVar <- round(100 * attr(pcaData, "percentVar")
```

9. The PCA output can be visualized using the R package ggplot2 [14], and saved as a PDF file.

```
library(ggplot2)
gg= ggplot(pcaData, aes(PC1, PC2, color=Group)) +
 geom_point(size=3) +
 xlab(paste0("PC1: ",percentVar[1],"% variance")) +
 ylab(paste0("PC2: ",percentVar[2],"% variance")) +
 coord_fixed()
ggsave("pca.pdf", plot=gg)
```

4 Notes

1. The FASTQ reads may contain low-quality bases, reads, or adapter sequences that need to be cleaned. Standalone tools like FastQC [15] can sample the FASTQ files and output visual representation of different FASTQ quality parameters. Other tools, like FASTX-Toolkit [16] and Trimmomatic [17], can then be used to trim the FASTQ reads.
2. An RNA-seq analysis pipeline consists of multiple steps, starting from read alignment to the reference genome, to identification of differentially expressed genes. The numerous tools available for the different steps can be combined to generate different data analysis pipelines. Moreover, new programs that improve upon existing algorithms are being constantly developed. Therefore, a thorough literature search about the tools and their intended purpose would help customize an existing pipeline or build a new pipeline.
3. The above pipeline built using HISAT2 [8], StringTie [9, 18], and DESeq2 [10] is a reference genome-based transcript mapping analysis pipeline. So, the first requirement for this pipeline is the availability of a sequenced genome.
4. If the reference genome is well-annotated, splice-site and exon information can be extracted from the annotation files using two scripts, `extract_splice_sites.py` and `extract_exons.py`, provided with HISAT2. These files can then be used as input when building the reference genome index, as follows:

```
hisat2-build -p 8 --ss genome.ss --exon genome.exon genome.fa
genome
```

   A reference genome annotation (gff file) can also be used to guide the transcript assembly and merge steps in StringTie.
5. RNA-seq data is generated from mRNA molecules where introns have been spliced out. So, when the RNA-seq reads are being aligned to a reference genome, a splice-aware aligner, like HISAT2 (**H**ierarchical **I**ndexing for **S**pliced **A**lignments of **T**ranscripts), needs to be used. Two additional commonly used splice-aware aligners are TopHat2 [19] and STAR [20].
6. The BAM files containing the HISAT2 [8] mapping results should be indexed using SAMtools [11] for faster access.
7. Two additional alignment strategies for RNA-seq analysis are de novo transcript assembly and read alignment to transcriptome. When a reference genome is not available, transcripts can be assembled from RNA-seq reads using tools like Trinity [21], Velvet [22], and SPAdes [23]. The second approach of aligning the reads to the transcriptome works for read datasets where read length is less than 50 bp.

8. Each program used in this pipeline has several additional parameters that were not discussed in this chapter. Users should consult the program manuals to figure out the full range of tasks a program can accomplish and if the use of additional flags may help with their analyses.
9. Different treatment contrasts for differential gene expression are set up using the design or model matrix in the DESeq2 analysis. The design matrix needs to be set up during or after loading the dataset as a DESeq object, i.e., **step 4** of Subheading 3.2, and before DESeq analysis is run, i.e., **step 5** of Subheading 3.2. In the current chapter we describe a "one factor (`"Group"`) with two levels (`"Control"` and `"Treatment"`)" design output. There is a flexibility to set up complex designs with two or three factors/levels, or for experiments with nested/imbalanced design.
10. The default output from DESeq2 [10] analysis is a seven-column text file, with the following information, namely, gene ID, baseMean, log2FoldChange, lfcSE, stat, p-value, and p-adj. However, DESeq2 provides an option to access the normalized counts, as shared above.
11. The DESeq2 [10] output generated above lists results for all the transcripts. In order to identify the significant differentially expressed transcripts, the dataset can be filtered based on the p-adj and log2 fold change columns. Generally, a p-adj value of 0.05 and log2 fold change cut-off of 2 is used.
12. The end product of this analysis pipeline is a list of up- and downregulated transcripts. Further work is needed to put this data into the context of the experiment conducted, and the hypothesis being tested. Gathering additional information, like putative gene function, expression in other experiments, co-regulation with other genes, and their role in different metabolic pathways, provides supporting evidence to build the narrative.
13. There is no consensus in the literature on the cut-off values. These vary depending on the experiment and relevance of different genes to a given pathway.

Acknowledgments

This work is supported by the USDA National Institute of Food and Agriculture, Hatch project FLA-FTL-005926. Any opinions, findings, conclusions, or recommendations expressed in this publication are those of the author(s) and do not necessarily reflect the view of the National Institute of Food and Agriculture (NIFA) or the US Department of Agriculture (USDA).

References

1. Van den Berge K, Hembach KM, Soneson C, Tiberi S, Clement L, Love MI, Patro R, Robinson MD (2019) RNA sequencing data: Hitchhiker's guide to expression analysis. Annu Rev Biomed Data Sci 2:139–173
2. Besemer J, Borodovsky M (1999) Heuristic approach to deriving models for gene finding. Nucl Acid Res 27:3911–3920
3. Allen JE, Salzberg SL (2005) JIGSAW: integration of multiple sources of evidence for gene prediction. Bioinformatics 21:3596–3603
4. Goldstein LD, Cao Y, Pau G, Lawrence M, Wu TD, Seshagiri S, Gentleman R (2016) Prediction and quantification of splice events from RNA-Seq data. PLoS One 11(5):e0156132
5. Elliott ML, Des Jardin EA, O'Donnell K, Geiser DM, Harrison NA, Broschat TK (2010) *Fusarium oxysporum* f. sp. *palmarum*, a novel forma specialis causing a lethal disease of *Syagrus romanzoffiana* and *Washingtonia robusta* in Florida. Plant Dis 94:31–38
6. Mortazavi A, Williams BA, McCue K, Schaeffer L, Wold B (2008) Mapping and quantifying mammalian transcriptomes by RNA-Seq. Nat Methods 5:621–628
7. Costa-Silva J, Domingues D, Lopes FM (2017) RNA-Seq differential expression analysis: an extended review and a software tool. PLoS One 12:e0190152
8. Kim D, Paggi JM, Park C, Bennett C, Salzberg SL (2019) Graph-based genome alignment and genotyping with HISAT2 and HISAT-genotype. Nat Biotechnol 37:907–915
9. Pertea M, Pertea GM, Antonescu CM, Chang T-C, Mendell JT, Salzberg SL (2015) StringTie enables improved reconstruction of a transcriptome from RNA-seq reads. Nat Biotechnol 33:290–295
10. Love MI, Huber W, Anders S (2014) Moderated estimation of fold change and dispersion for RNA-seq data with DESeq2. Genome Biol 15:550
11. Li H, Handsaker B, Wysoker A, Fennell T, Ruan J, Homer N, Marth G, Abecasis G, Durbin R, 1000 Genome Project Data Processing Subgroup (2009) The sequence alignment/map (SAM) format and SAMtools. Bioinformatics 25:2078–2079
12. R Core Team (2020) R: a language and environment for statistical computing. R Foundation for Statistical Computing, Vienna. https://www.R-project.org/
13. Huber W et al (2015) Orchestrating high-throughput genomic analysis with Bioconductor. Nat Methods 12:115–121
14. Wickham H (2016) ggplot2: elegant graphics for data analysis. Springer, New York
15. Andrews S (2010) FastQC: a quality control tool for high throughput sequence data. Available online at: http://www.bioinformatics.babraham.ac.uk/projects/fastqc
16. Hannon GJ (2010) FASTX-Toolkit FASTQ/A short-reads pre-processing tools. Available online at: http://hannonlab.cshl.edu/fastx_toolkit
17. Bolger AM, Lohse M, Usadel B (2014) Trimmomatic: a flexible trimmer for Illumina sequence data. Bioinformatics 30:2114–2120
18. Pertea M, Kim D, Pertea GM, Leek JT, Salzberg SL (2016) Transcript-level expression analysis of RNA-seq experiments with HISAT, StringTie and Ballgown. Nature Protoc 11:1650–1667
19. Kim D, Pertea G, Trapnell C, Pimentel H, Kelley R, Salzberg SL (2013) TopHat2: accurate alignment of transcriptomes in the presence of insertions, deletions and gene fusions. Genome Biol 14:R36
20. Dobin A, Davis CA, Schlesinger F, Drenkow J, Zaleski C, Jha S, Batut P, Chaisson M, Gingeras TR (2013) STAR: ultrafast universal RNA-seq aligner. Bioinformatics 29:15–21
21. Haas BJ, Papanicolaou A, Yassour M, Grabherr M, Blood PD, Bowden J, Couger MB, Eccles D, Li B, Lieber M, Macmanes MD, Ott M, Orvis J, Pochet N, Strozzi F, Weeks N, Westerman R, William T, Dewey CN, Henschel R, Leduc RD, Friedman N, Regev A (2013) De novo transcript sequence reconstruction from RNA-seq using the trinity platform for reference generation and analysis. Nat Protoc 8:1494–1512
22. Zerbino DR, Birney E (2008) Velvet: algorithms for de novo short read assembly using de Bruijn graphs. Genome Res 18:821–829
23. Bushmanova E, Antipov D, Lapidus A, Prjibelski AD (2019) rnaSPAdes: a de novo transcriptome assembler and its application to RNA-Seq data. GigaScience 8:giz100

Chapter 5

Quantification and Isolation of Spontaneous Colony Growth Variants

Cristina López Díaz, Lucía Gómez-Gil, Elena Pérez-Nadales, Gesabel Navarro Velasco, and Antonio Di Pietro

Abstract

The appearance of colony growth sectors on solid medium plates has been described in many fungi. Although the molecular bases of this phenomenon remain largely unknown, possible relationships with genetic or epigenetic changes have been reported. Here we present a method to quantify the frequency of colony growth sectors in *Fusarium oxysporum*, which can be used to compare different fungal strains and to infer their genetic instability.

Key words Growth variant, Colony sectorization, Phenotypic plasticity

1 Introduction

The long-term survival of organisms depends on their ability to adapt to environmental changes through genetic variation [1]. In fungi, such adaptive capacity is often reflected by phenotypic growth differences on culture media plates, which can be readily observed as colony sectors [2, 3]. These growth phenotypes are determined by various factors such as the composition of the medium, the age of the colony, or the incubation temperature [4]. This phenomenon known as "sectorization" causes changes between the predominant colony phenotype and the "variant" phenotype. The underlying mechanisms of this sector-shaped phenotypic plasticity are still poorly understood, although several studies suggest it may have both genetic and epigenetic origins. In *Fusarium* and *Metarhizium*, the formation of colony sectors was suggested to be associated to epigenetic factors such as changes in DNA methylation, which also lead to a reduction in fungal virulence [5, 6]. In *Podospora anserina*, the appearance of differential growth sectors exhibiting epigenetic cell degeneration known as

Jeffrey Coleman (ed.), *Fusarium wilt: Methods and Protocols*,
Methods in Molecular Biology, vol. 2391, https://doi.org/10.1007/978-1-0716-1795-3_5,

"Crippled Growth" was shown to be under control of both genetic and environmental factors [7, 8].

We routinely observe the emergence of differentiated colony growth sectors in the fungal pathogen *Fusarium oxysporum*. While the exact mechanisms underlying this phenomenon remain to be elucidated, we found that it provides a useful tool for analyzing phenotypic plasticity in this species complex, which includes numerous clonal lineages and *formae speciales*, where genome analysis revealed differences in chromosome number and organization [9]. In this chapter we describe an optimized method which allows the systematic analysis and efficient quantification of growth sectors in multiple colonies of *F. oxysporum*.

2 Materials

Prepare all solutions using ultrapure water (prepared by purifying deionized water, to attain a sensitivity of 18.2 MΩ-cm at 25 °C). Prepare and store all reagents at room temperature unless indicated otherwise. Diligently follow all waste disposal regulations when disposing waste materials.

1. Puhalla's Minimal Medium (PMM) [10] (*see* **Note 1**): Add 30 g sucrose, 1 g KH_2PO_4, 0.5 g KCl, 0.5 g $MgSO_4 \times 7\ H_2O$, and 15 agar to 600 mL water. Adjust to final volume of 1 L (*see* **Note 2**) (Fig. 1).
2. 375 mM Methionine solution: Add 80 mL of water to a glass beaker and dissolve 5.6 g of L-Methionine, and adjust the pH to 2.5 with a dilution of HCl (*see* **Note 3**). Adjust to 100 mL and sterilize by filtration through a 0.22 μm pore size filter. Store at 4 °C.
3. Other nitrogen sources. Prepare concentrated stock solutions of the different nitrogen sources at the following concentrations: 240 mM sodium nitrate; 120 mM ammonium sulfate; 120 mM urea; 120 mM hydroxyurea; 120 mM glutamine; and 240 mM glutamic acid. Adjust to 100 mL and sterilize by filtration through a 0.22 μm pore size filter. Store at 4 °C.
4. 10 N NaOH solution: Dissolve 40 g NaOH in 80 mL of water by stirring. Since this is an exothermic reaction, it is recommended to add the NaOH pearls little by little. Adjust to 100 mL.
5. 2.4 N HCl solution.
6. Potato dextrose broth (PDB) medium: boil 200 g of peeled potatoes in 0.6 L of deionized water for 1 h. Stir and add 20 g of glucose (2%) (w/v). Add deionized water up to 1.0 L. Autoclave.

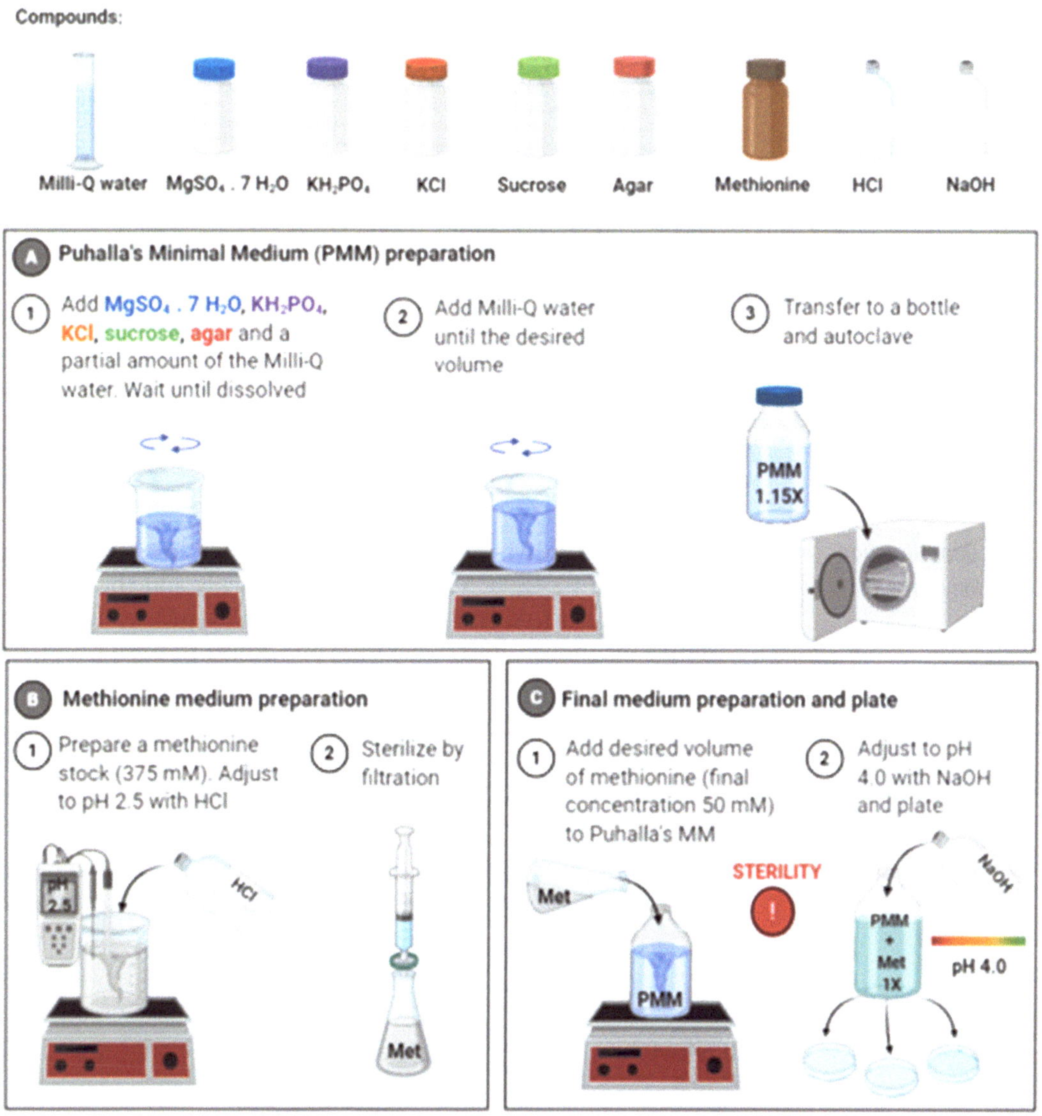

Fig. 1 Preparation of Puhalla's Minimal Medium (PMM) supplemented with methionine. (**a**) Weigh the different ingredients of PMM, adjust with water to the desired final volume to obtain a 1.15× solution, and autoclave. (**b**) Prepare a 375 mM methionine stock solution in water, adjust the pH to 2.5 with 2.4 M HCl for solubilization, and filter-sterilize. (**c**) Add the appropriate volume of methionine stock to the autoclaved PMM cooled to 45 °C, adjust pH to 4.0 with 10 N NaOH, and pour into Petri dishes

7. Potato dextrose agar (PDA). Dissolve 39 g potato dextrose agar in water and adjust to 1 L. Autoclave and pour to petri plates.

3 Methods

Carry out all procedures under sterile conditions at room temperature, unless otherwise stated.

Various compounds were tested as potential growth restrictors to optimize the detection and quantification of sectors. Finally, Puhalla's Minimal Medium supplemented with methionine was used since this amino acid restricts the growth of the colony, allowing a more accurate quantification of the sectors generated (Fig. 2).

3.1 Quantification of Colony Growth Sectors

1. To prepare PMM with methionine, let the autoclaved PMM cool down to 45 °C (*see* **Note 4**). Gradually add the methionine solution to a final concentration of 50 mM while mixing with a magnetic stirrer. Check pH and adjust to 4.0 if necessary and quickly pour the medium into Petri dishes and let solidify.
2. Inoculate 2×10^6 microconidia of the desired *F. oxysporum* strain in a 125 mL Erlenmeyer flask containing 25 mL PDB medium. Incubate for 3 days at 28 °C with orbital shaking at 170 rpm.
3. Collect microconidia by filtration through an autoclaved nylon filter (mesh size 10 μm) or cheesecloth and centrifuge 10 min

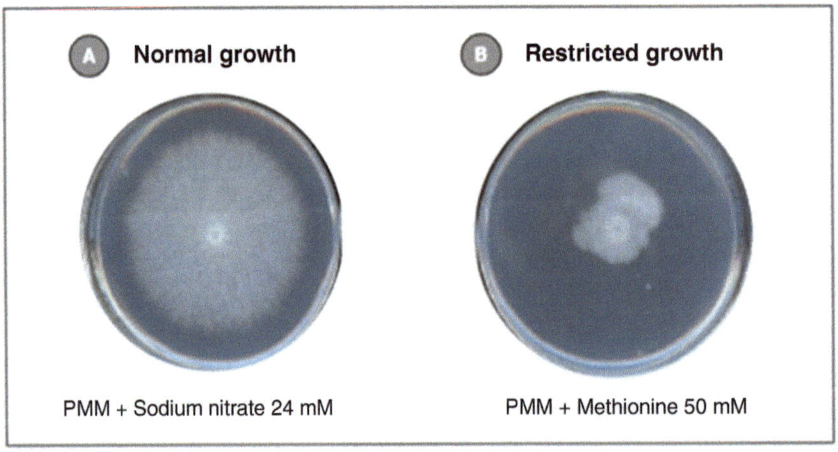

Fig. 2 Generation of colony growth sectors by *Fusarium oxysporum. F. oxysporum* strain Fol4287 was grown for 7 days on Puhalla's Minimal Medium (PMM) supplemented with sodium nitrate (**a**) or methionine as the nitrogen source (**b**). Note the restricted colony growth and the appearance of growth sectors on methionine medium

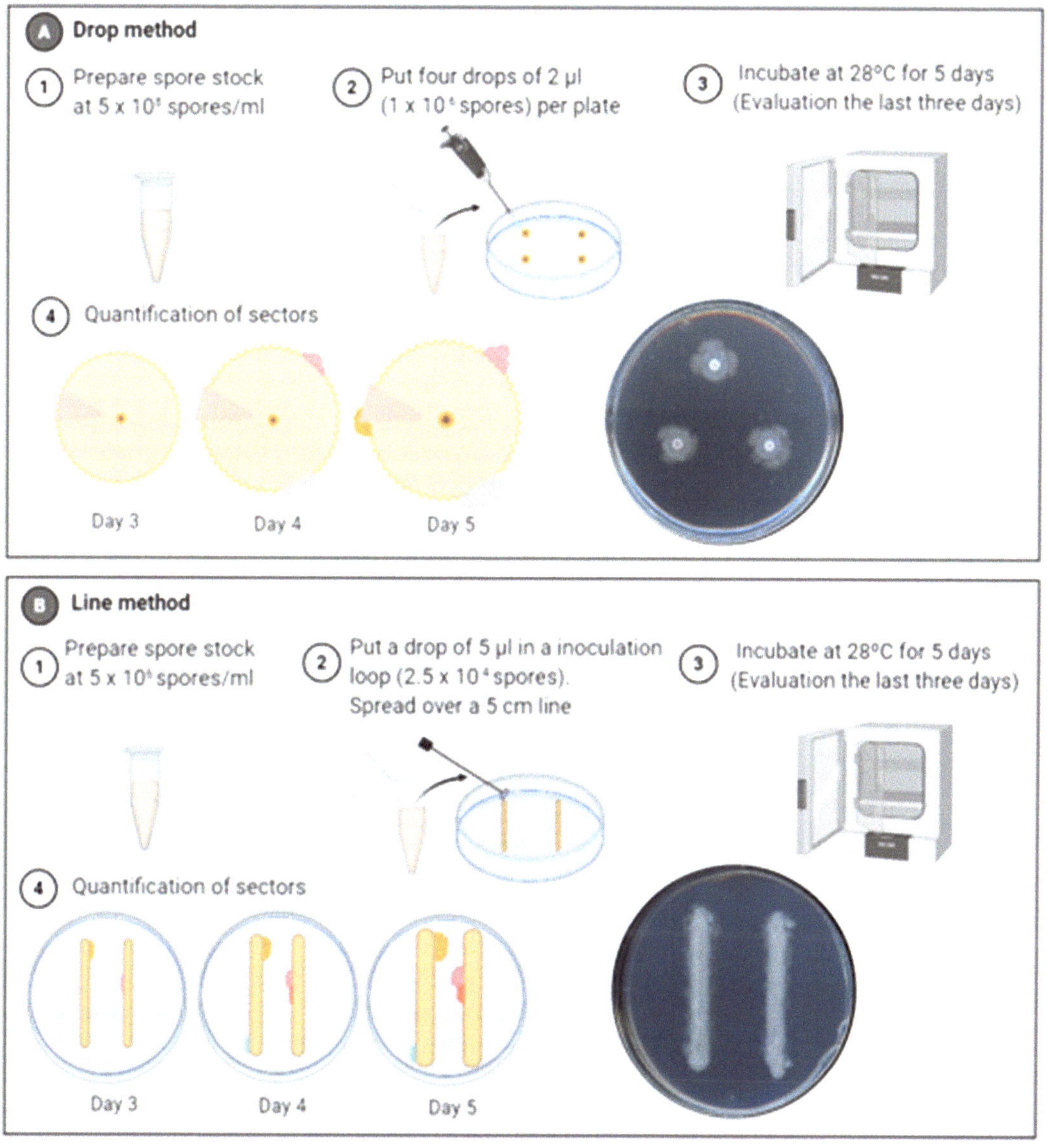

Fig. 3 Quantification of colony growth sectors. (**a**) Drop method: Spot-inoculate 3 to 4 aliquots of 2 μL containing 1×10^6 microconidia on a PMM + methionine plate. Incubate at 28 °C count the emerging colony growth sectors starting on day 3 until day 5. (**b**) Line method: Place a 5 μL drop containing 2.5×10^4 microconidia in an inoculation loop and spread along a 5 cm line (2 lines per plate). Incubate at 28 °C and count the emerging colony growth sectors starting at day 3 until day 5. Perform at least 6 replicates for both methods

at 2236 × *g*. Resuspend microconidia in 1 mL of water and determine concentration in a hemocytometer.

4. For quantification of colony growth sectors, two different methods can be used (Fig. 3).

Drop method: Adjust microconidia concentration to 5×10^8/mL. Spot-inoculate 2 μL aliquots at three or four equidistant positions on the PMM plate.

Line method: Adjust microconidia concentration to 5×10^6/mL. Add a drop of 5 μL to a bacterial loop, place the loop carefully on the surface of the PMM plate, and spread the conidia along a line of 5 cm (*see* **Note 5**).

For both methods, incubate plates for 5 days at 28 °C in the dark.

5. Sectors of growth variants are counted on days 3, 4, and 5 after inoculation (*see* **Note 6**).

3.2 Isolation and Phenotypic Characterization of Colony Sectors

1. To prepare PMM with different nitrogen sources, let the autoclaved PMM cool down to 60 °C. Add the appropriate volume of each nitrogen source stock to obtain the following final concentrations: 24 mM sodium nitrate; 12 mM ammonium sulfate; 12 mM urea; 12 mM hydroxyurea; 12 mM glutamine; and 24 mM glutamic acid. Quickly pour the media into Petri dishes and let solidify.
2. Adjust microconidia concentration to 5×10^8/mL. Spot-inoculate a 5 μL drop in the center of the plate and let dry in a sterile hood. Incubate plates at 28 °C for 7 days.
3. Sectors usually have a triangular shape and display faster growth than the wild-type strain (*see* **Note 7**). For sector isolation, take a small piece of agar from the growing edge of the sector using a sterile toothpick, and transfer to a new plate of the same medium.
4. Inoculate growth variants on PMM containing either glutamic acid or methionine and on PDA plates to characterize the growth phenotype. Incubate for 6–7 days, and score the colony growth phenotype with respect to the original strain (Fig. 4).

4 Notes

1. The basic medium used here lacks nitrate (as described in the original protocol) and instead uses different nitrogen sources.
2. To calculate the final volume of water, take into account the volume of the given nitrogen solution which will be added after autoclaving.
3. To increase the solubility of methionine in water, the pH of the stock solution should be 2.5.
4. Methionine is degraded if added to very hot medium.
5. Before spreading the conidia, draw a 5 cm line with a permanent ink pen on the reverse side of the plate.

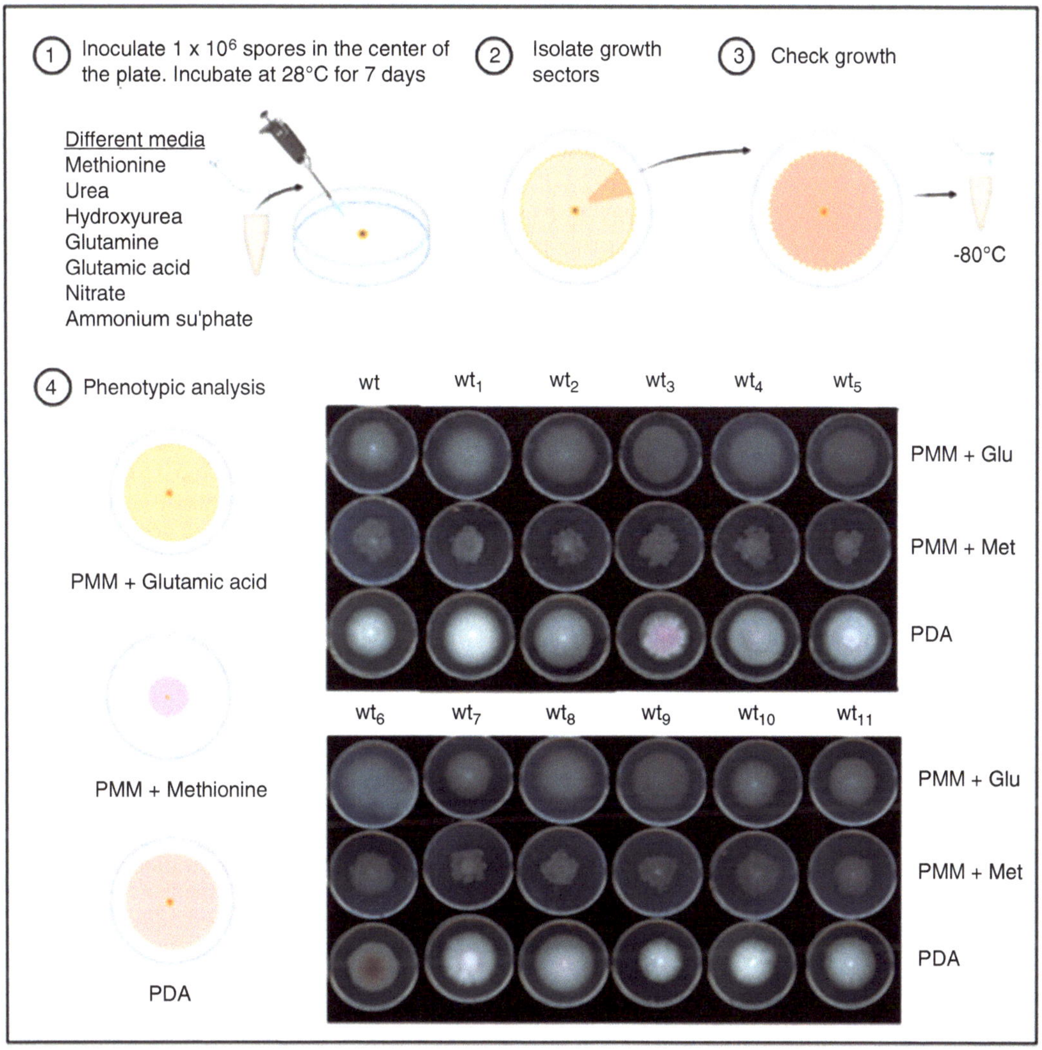

Fig. 4 Isolation and phenotypic analysis of colony growth sectors. Spot-inoculate 1×10^6 microconidia of the wild-type strain on PMM plates with different nitrogen sources. After 7 days at 28 °C, colony sectors showing faster growth are isolated, transferred to three different media (PMM + glutamic acid; PMM + methionine; PDA), and classified according to their differential growth phenotypes

6. As growth variants, we consider sectors that exhibit a higher growth speed compared to the wild type. Because sectors that emerge at different time points may overlap, each sector should be marked on the back of the plate after scoring in order to avoid double counting. We also recommend using different marker colors for each scoring day.
7. On some plates no sectors may be observed, depending on the nitrogen source and the fungal strain used.

Acknowledgments

The authors acknowledge Esther Martinez Aguilera for valuable technical help. Figures were prepared using the free Biorender online tool. Research in the Di Pietro lab is supported grants BIO2016-78923-R and PID2019-108045RB-I00 from the Spanish Ministerio de Ciencia e Innovación (MICINN). L.G.G is supported by a FPU PhD fellowship from the Spanish Ministerio de Educacion y Ciencia (MEC).

References

1. Friedman G, McCarthy S, Rachinskii D (2014) Hysteresis can grant fitness in stochastically varying environment. PLoS One 9:e103241
2. Stevens FL (1922) The *Helminthosporium* foot-rot of wheat: with observations on the morphology of *Helminthosporium* and on the occurrence of saltation in the genus—II. Evidence and discussion of the occurrence of saltation within the genus *Helminthosporium*. Illinois Nat Hist Surv Bull 14:139–178. plates XXII–XXXII
3. Silar P (2019) Phenotypic instability in fungi. Adv Appl Microbiol 107:141–187
4. Slepecky RA, Starmer WT (2009) Phenotypic plasticity in fungi: a review with observations on *Aureobasidium pullulon*. Mycologia 101:823–832
5. Kim DH (1997) Induced change in DNA methylation of *Fusarium oxysporum* f. sp. *niveum* due to successive transfer. J Biochem Mol Biol 30:216–221
6. Wang C, Butt TM, Leger RJ (2005) Colony sectorization of *Metarhizium anisopliae* is a sign of ageing. Microbiology 151:3223–3236
7. Haedens V, Malagnac F, Silar P (2005) Genetic control of an epigenetic cell degeneration syndrome in *Podospora anserina*. Fungal Genet Biol 42:564–577
8. Gautier V, Tong LCH, Nguyen T-S, Debuchy R, Silar P (2018) *PaPro1* and *IDC4*, two genes controlling stationary phase, sexual development and cell degeneration in *Podospora anserina*. J Fungi 4:85
9. Ma LJ, van der Does HC et al (2010) Comparative genomics reveals mobile pathogenicity chromosomes in *Fusarium*. Nature 464:367–373
10. Puhalla JE (1985) Classification of strains of *Fusarium oxysporum* on the basis of vegetative compatibility. Can J Bot 63:179–171

Chapter 6

Agrobacterium tumefaciens-Mediated Transformation Method for *Fusarium oxysporum*

Yan-Hong Dong and Shu-Tong Wang

Abstract

Agrobacterium tumefaciens-mediated transformation (ATMT) is becoming a popular effective system as an insertional mutagenesis tool in filamentous fungi. An efficient *Agrobacterium tumefaciens*-mediated transformation approach was developed for the plant pathogenic fungus, *F. oxysporum*, the causal agent of Apple replant disease (ARD) in China. Four parameters were selected to optimize efficiencies of transformation. *A. tumefaciens* concentration, conidial concentration of *F. oxysporum*, and co-culture temperature and time have a significant influence on all parameters. Transformants emit green fluorescence under fluorescence microscopy. The integration of a mitotically stable hygromycin resistance gene (hph) in the genome is confirmed by PCR. The transformation efficiency can reach up to 300 transformants per 10^6 conidia under optimal conditions. This ATMT method is an efficient tool for insertional mutagenesis of *F. oxysporum*.

Key words *Fusarium oxysporum*, *Agrobacterium tumefaciens-mediated transformation*, T-DNA, Optimization

1 Introduction

Various DNA-mediated transformation systems have been developed and applied to different filamentous fungi for genetic research in recent years. Traditional tools, such as PEG/$CaCl_2$-protoplast transformation, restriction enzyme-mediated integration (REMI) [1], and genetic engineering by RNA interference (RNAi) [2], have been successfully developed for the construction of random mutant libraries in different fungal species [3, 4]. Compared with these methods, *Agrobacterium tumefaciens*-mediated transformation (ATMT) has several advantages. One of the principal advantages is the flexibility in choosing starting material (protoplasts, hyphae, spores, or mycelial tissues) to be transformed [1, 5–7]. Other advantages include high transformation efficiency [8], high rate of single-copy insertions, and easy identification of flanking sequences of T-DNA [9]. Therefore, ATMT was thought to be an

Jeffrey Coleman (ed.), *Fusarium wilt: Methods and Protocols*,
Methods in Molecular Biology, vol. 2391, https://doi.org/10.1007/978-1-0716-1795-3_6,

effective tool to explore pathogenicity-related genes [10–12]. *A. tumefaciens*-mediated transformation of filamentous fungi has been widely reported since 1998 [13].

To date, more than 200 fungal species have been transformed by ATMT [14–17]. Here, we introduce a way to establish and optimize an ATMT system to construct a mutant library of *F. oxysporum*.

2 Materials

2.1 Biological Organisms

1. *F. oxysporum* isolate to mutate, in this instance HS2.
2. *Agrobacterium tumefaciens* strain LBA4404, carrying pCam-hybgfp vector.

2.2 Agrobacterium-Mediated Transformation

1. LB liquid medium: Weigh 8.0 g tryptone, 5.0 g yeast extract, and 10.0 g NaCl, and bring to 1 L with water. Sterilize by autoclaving at 121 °C for 20 min. Store at room temperature.
2. Potato Dextrose Agar: Weigh 200.0 g potatoes, 15.0 g agar, and 20.0 g glucose, and bring to 1 L with water. Sterilize by autoclaving at 121 °C for 20 min. Store at room temperature.
3. Potato Dextrose Broth: Weigh 200.0 g potatoes and 20.0 g glucose, and bring to 1 L with water. Sterilize by autoclaving at 121 °C for 20 min. Store at room temperature.
4. 1% $CaCl_2 \cdot 2\ H_2O$: Dissolve 10 g $CaCl_2 \cdot 2\ H_2O$ in 1 L of water. Sterilize by autoclaving at 121 °C for 20 min. Store at 4 °C.
5. 0.01% $FeSO_4 \cdot 7\ H_2O$: Weigh 0.1 g $FeSO_4 \cdot 7\ H_2O$ and dissolve in 1 L of sterile water, and filter through a 0.45 μm water-based filter. Store at 4 °C.
6. 20% Glucose: Dissolve 200.0 g of glucose in 1 L of sterile water, and filter through a 0.45 μm water-based filter. Store at 4 °C.
7. 50% glycerol: Add 100 mL glycerol to 100.0 mL water. Sterilize by autoclaving at 121 °C for 20 min, and store at 4 °C.
8. 1.25 M KH_2PO_4: Dissolve 17.0 g KH_2PO_4 in 100.0 mL of water. Sterilize by autoclaving at 121 °C for 20 min, and store at 4 °C.
9. 1.25 M K_2HPO_4: Dissolve 21.8 g K_2HPO_4 in 100.0 mL of water. Sterilize by autoclaving at 121 °C for 20 min, and store at 4 °C.
10. K-buffer: Mix well equal volumes of 1.25 M KH_2PO_4 and 1.25 M K_2HPO_4. Adjust the pH to 4.8 and autoclave at 121 °C for 20 min, and store at 4 °C.

11. 1 M $MgSO_4·7\ H_2O$: Dissolve 123.24 g of $MgSO_4·7\ H_2O$ in 500 mL water. Sterilize by autoclaving at 121 °C for 20 min, and store at 4 °C.
12. MN-buffer: Dissolve 30.0 g $MgSO_4·7\ H_2O$ and 15.0 g NaCl in 1 L of water. Sterilize by autoclaving at 121 °C for 20 min, and store at 4 °C.
13. 20% NH_4NO_3: Dissolve 200.0 g NH_4NO_3 in 100 mL of water, and bring the final volume to 1 L with water. Sterilize by autoclaving at 121 °C for 20 min, and store at 4 °C.
14. 1 M MES: Dissolve 19.53 g of MES in 100.0 mL of water, and adjust pH 5.5 with NaOH. Filter through a 0.45 μm water-based filter. Store at −20 °C.
15. Inducing medium (liquid): Mix 0.8 mL K-buffer, 20 mL MN-buffer, 1 mL 1% $CaCl_2·2\ H_2O$, 10 mL 0.01% $FeSO_4·7\ H_2O$, 5 mL 1 M Tris, 2.5 mL 20% NH_4NO_3,10 mL 50% glycerol, 10 mL 20% glucose, 40 mL 1 M MES (pH 5.5), and bring to a final volume of 1 L with water. Sterilize by autoclaving at 121 °C, 20 min. Store at 4 °C.
16. 0.2 mol/mL acetosyringone (AS): Add 785 mg acetosyringone in dimethyl sulfoxide to make a final volume of 20 mL. Filter with a 0.22 μm oil-based filter, and store at −20 °C in the dark.
17. 10 mg/mL rifampicin (Rif): Weigh 0.25 mg of rifampicin and dissolve in 25 mL dimethyl sulfoxide and filter with a 0.22 μm oil filter, and store at −20 °C in the dark.
18. 50 mg/mL kanamycin (Kan): Dissolve 1 g of kanamycin in 20 mL water, and filter with a 0.45 μm water-based filter and store at −20 °C.
19. 100 mg/mL chloramphenicol (CM): Dissolve 1 g of chloramphenicol in 10 mL dimethyl sulfoxide until completely dissolved, and filter through a 0.22 μm oil filter and store at −20 °C.
20. 100 mg/mL ampicillin (AMP): Mix 1 g of ampicillin in 10 mL of sterile water until completely dissolved, and sterilize with a 0.45 μm water-based filter and store at −20 °C.
21. 100 mg/mL streptomycin (Str): Mix 1 g of streptomycin in 10 mL of sterile water until completely dissolved, and sterilize with a 0.45 μm water-based filter and store at −20 °C.
22. 100 mg/mL hygromycin B (HygB): Weigh 1 g of hygromycin and dissolve in 10 mL of sterile water. Sterilize with a 0.45 μm water-based filter and store at −20 °C in the dark.

2.3 Screening and Validation of Transformants

1. Fluorescence microscope with at least a 40× objective lens and filters for visualization of GFP.
2. Sterile toothpicks.
3. Liquid nitrogen.
4. Sterile mortar and pestle.
5. Commercial DNA extraction kit.
6. Reagents for PCR, i.e., PCR master mix, or buffer, dNTPs, $MgCl_2$, molecular grade water.
7. Primer HYF-F (5′- ATGAAAAAGCCTGAACTCAC-3′).
8. Primer HYG-R (5′- CTATTCCTTTGCCCTCG-3′).
9. Thermocycler.
10. Agarose gel and electrophoresis equipment.

3 Methods

3.1 Assessment of Isolate Sensitivity to Hygromycin B

1. Streak the wild-type *F. oxysporum* HS2 isolate from a glycerol stock, and allow to grow for 5 days on PDA plates at 25 °C in the dark.
2. Harvest the conidia of *F. oxysporum* HS2 by washing the cultures on the PDA media gently with a sterile solution of 0.05% Tween 80. Filter the conidial suspension with four layers of sterile gauze to remove mycelia fragments, and wash twice with sterile deionized water. Suspend the conidia in sterile deionized water, and adjust the concentration to 10^6 conidia/mL using a hemocytometer.
3. Mycelial plugs (d = 6 mm) are transplanted to PDA plates supplemented with different hygromycin B concentrations (0, 12.5, 25, 50, and 100 μg/mL). After cultivation at 25 °C for 7 days, colony diameters are measured to evaluate the minimum inhibition (Fig. 1) (*see* **Note 1**). In addition, 200 μL of the conidial suspension is spread with a sterile glass rod on PDA plates with different concentrations of hygromycin B (0, 12.5, 25, 50, and 100 μg/mL), and incubated at 25 °C for 18 h (Fig. 1).

3.2 A. tumefaciens-Mediated Transformation of F. oxysporum

The transformation procedure is based on the protocol of Mullins [17].

1. *A. tumefaciens* strain LBA4404 carrying the pCamhybgfp vector is grown on a LB agar plate supplemented with 50 μg/mL kanamycin and 100 μg/mL streptomycin at 28 °C for 2 days.
2. A single colony is picked from the LB agar plate, transferred in LB broth containing 50 μg/mL kanamycin and 100 μg/mL streptomycin, and incubated at 28 °C under agitation (220 rpm) for 24 h or until the OD_{600} reaches about 0.6.

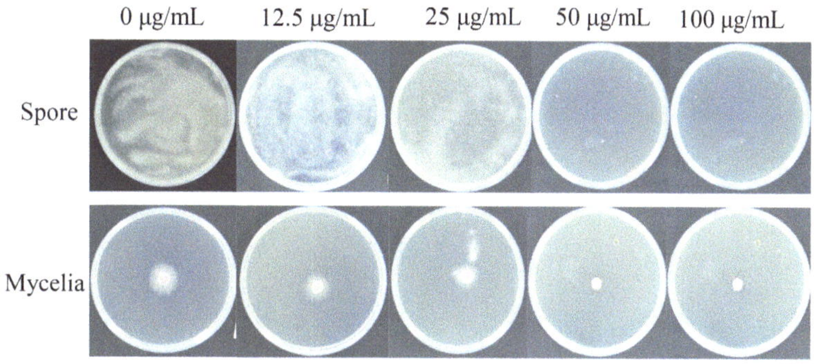

Fig. 1 The effect of different concentrations of hygromycin B on the mycelial growth and conidial germination of *F. oxysporum*

3. The *A. tumefaciens* culture is centrifuged at 5000 × *g* for 5 min at room temperature, washed twice with 1 mL of induction medium (IM) (liquid), diluted in 10 mL of IM [14, 15] with different concentrations of AS, and incubated at 28 °C at 220 rpm for 4–6 h until the OD_{600} reaches 0.2–0.3 (*see* **Note 2**).
4. The conidial suspension is obtained according to the procedure described above (**Step 2** of Subheading 3.1), and aliquots of the conidial suspension are adjusted to a concentration of 10^5, 10^6, and 10^7 spores/mL. *A. tumefaciens* cells are diluted to an OD_{600} of 0.3 (*see* **Note 3**).
5. *A. tumefaciens* suspensions are mixed with an equal volume of conidial suspension of *F. oxysporum*.
6. The *Agrobacterium*-conidia mixture (200 μL per plate) is applied on sterile qualitative filter paper and overlaid on the surface of inducing medium (IM) plate (same as IM except that it contains 1.5% agar) (*see* **Note 4**).
7. The co-cultivation plates are incubated for 48 h at 19 °C.
8. The filter paper is removed and cut into strips and transferred onto PDA medium containing 100 μg/mL hygromycin B.
9. Individual putative transformants are isolated to a new PDA plate containing 100 μg/mL hygromycin B and allowed to grow (Fig. 2) (*see* **Note 5**).

3.3 Confirmation of Generation of Transformants

3.3.1 Observation of Green Fluorescence

1. GFP will be observed if the T-DNA randomly integrated into the genome of the *F. oxysporum* HS2. The green fluorescence of the mutant should be observed using a fluorescence microscope (we used an Olympus IX71, Japan) to confirm integration of the T-DNA in the transformants. The wavelength of the excited light is 460–550 nm, and the emission wavelength is 590 nm (Fig. 3) (*see* **Note 6**).

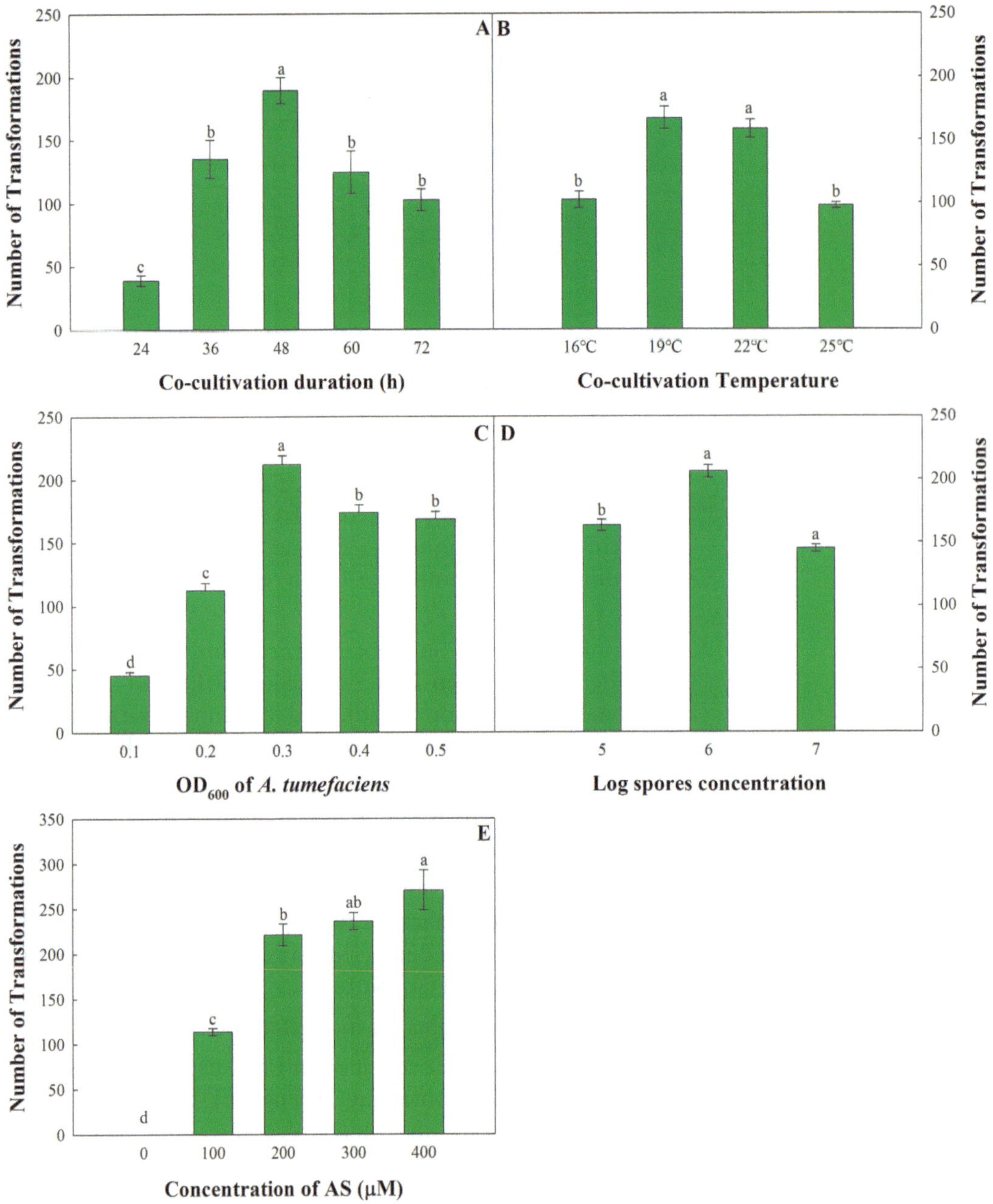

Fig. 2 Single-factor comparison of *A. tumefaciens*-mediated transformation of *F. oxysporum*. The factors include the cocultivation duration (**a**), the cocultivation temperature (**b**), the OD_{600} of *A. tumefaciens* (**c**), the conidia concentration of *F. oxysporum* (**d**), and the concentration of AS (**e**). Error bars represent standard deviation from repeats. Different letters indicate a significant difference between different treatments ($P < 0.05$)

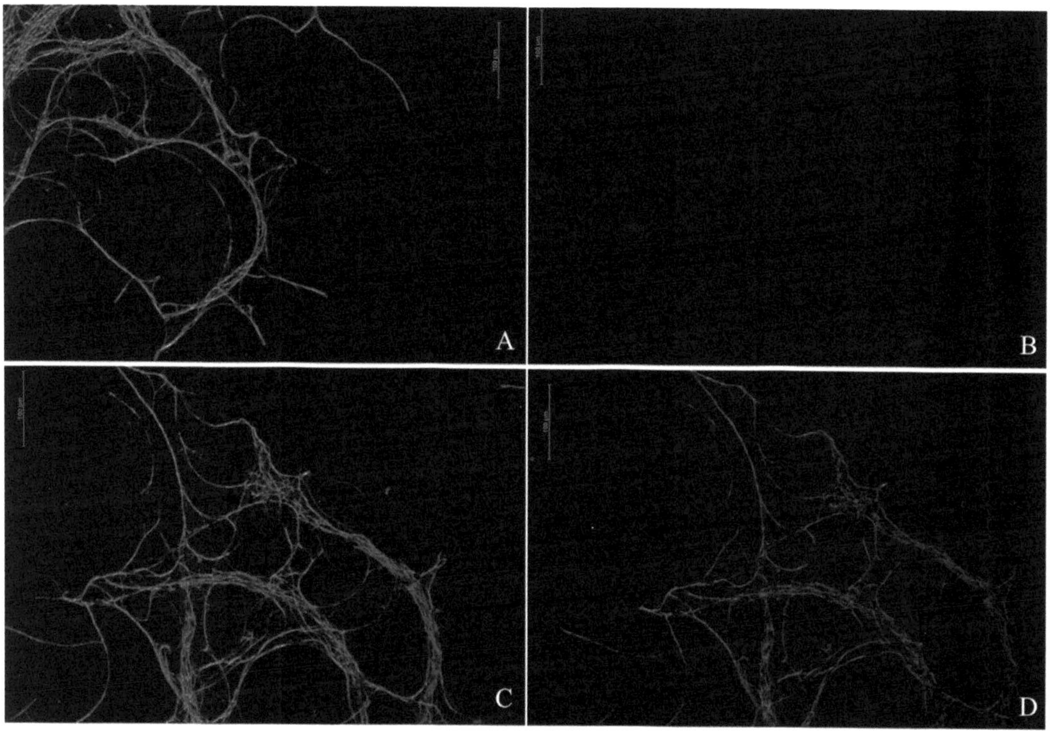

Fig. 3 Green fluorescent protein detection of the wild-type isolate, HS2, of *F. oxysporum* (**a**, **b**) and a transformant strain, HS2–2483 (**c**, **d**). (**a**, **c**) represent the bright field; (**b**, **d**) represent the fluorescent field

3.3.2 DNA Extraction and PCR Analyses

1. Transformants are inoculated into 50 mL of PDB containing 100 μg/mL hygromycin B and grown for 5 days at 25 °C with shaking at 160 rpm.
2. Harvest hyphae using sterilized toothpicks, and grind into a powder under liquid nitrogen using a sterile mortar and pestle.
3. Fungal genomic DNA is isolated using a commercially available fungal DNA extraction kit (*see* **Note 7**).
4. To detect the integration of T-DNA in the putative transformants, amplification by PCR should be carried out for the hygromycin phosphotransferase gene (*hph*) with the specific primers HYG-F and HYG-R.
5. The following procedure is used: 94 °C for 5 min, then 94 °C for 40 s, 51 °C for 40 s, and 72 °C for 2 min for 35 cycles, followed by 72 °C for 10 min [18].
6. The genomic DNA of the wild-type isolate, HS2, is used as a negative control, and the plasmid pCamhybgfp should be used as a positive control. The PCR products are detected using 1.0% agarose gel electrophoresis.
7. The positive transformation of T-DNA is confirmed by amplifying the *hph* gene using the genomic DNA of the transformants, which produces a 1000 bp fragment (Fig. 4).

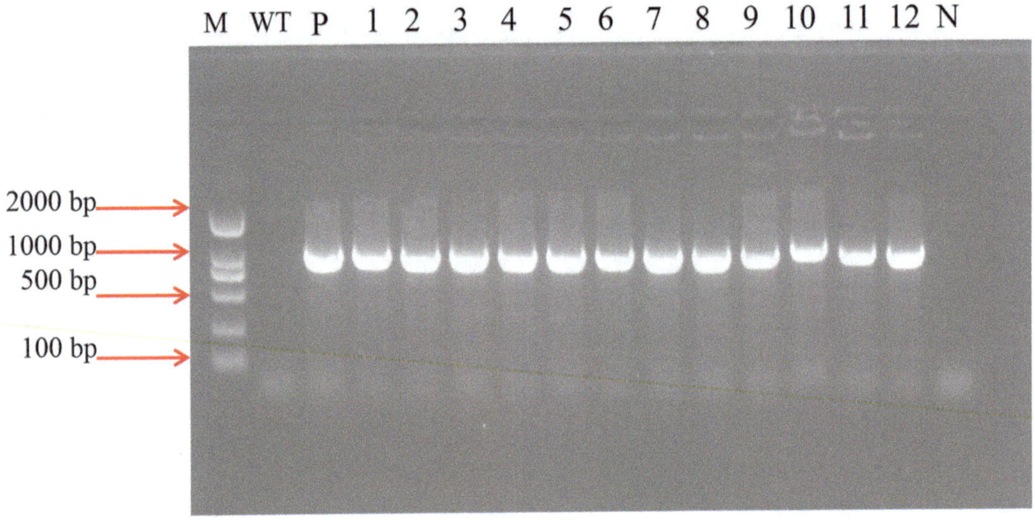

Fig. 4 PCR detection of the *hph* gene in transformants using primer pairs of HYG-F and HYG-R. Lane M represents the DNA Marker; lane P represents the positive control, pCamhybgfp; lanes 1 to 11 represent different transformants generated by ATMT; lane WT represents the wild-type strain of *F. oxysporum*, HS2; lane N represents the negative control

3.4 Mitotic Stability Test

1. To determine whether the insertion is genetically stable or not, transformants of *F. oxysporum* that contain the hph gene and the green fluorescence are randomly selected for subculturing on PDA with 100 μg/mL hygromycin B for 8 days.
2. Following colonization of each plate, a mycelial plug is taken from the edge of the colony and transferred onto a fresh PDA plate without hygromycin B and allowed to grow at 25 °C for 5 days, from which the new mycelia are placed onto a fresh PDA plate in the absence of hygromycin B.
3. After five generations, the mycelia are transferred on PDA plate containing 100 μg/mL hygromycin B to check the growth of mycelium (Fig. 5) (*see* **Note 8**).

4 Notes

1. An example of expected results is given in Fig. 1. The results indicate that the mycelial growth of *F. oxysporum* is partially inhibited by hygromycin B at a concentration of 50 μg/mL and is completely inhibited at a concentration of 100 μg/mL. Therefore, a final concentration of 100 μg/mL of hygromycin B in the PDA plates would be recommended as the optimal concentration to screen transformants.
2. The concentration of AS can be optimized based on each specific transformation system. To determine the optimal AS

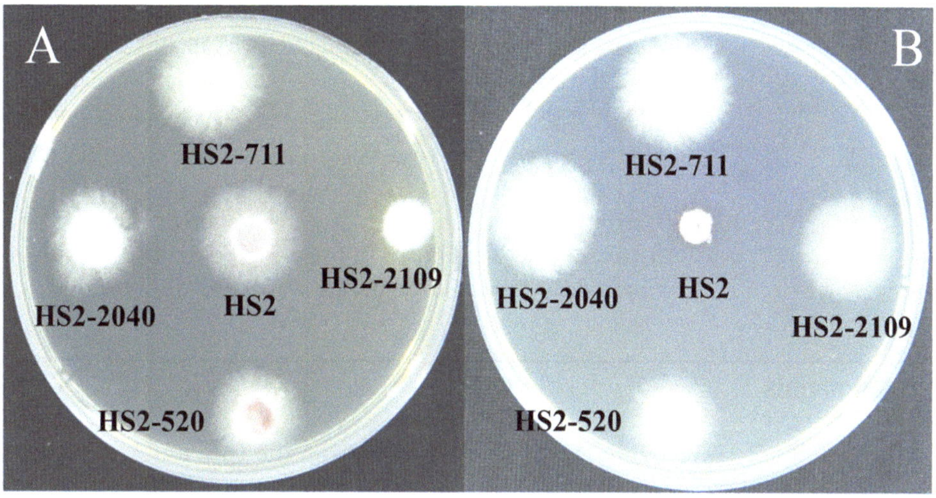

Fig. 5 Genetic stability of different transformants. Growth of wild-type strain HS2 and different transformants on PDA medium without hygromycin (**a**). Wild-type strain HS2 and four transformants after growth for five generations on PDA medium containing of 100 μg/mL hygromycin (**b**)

concentration, different concentrations (0, 100, 200, 300, and 400 μmol/L) can be added to the co-culture media. AS-free medium would be used as the control.

3. To obtain individual transformants and avoid transformants overlapping during growth, the concentration of the *A. tumefaciens* bacterial suspension (OD_{600} = 0.1, 0.2, 0.3, 0.4, and 0.5) and the concentration of the *F. oxysporum* conidia suspension can be optimized (10^5, 10^6, and 10^7spores/mL) for each individual isolate.
4. Optimization of co-culture temperature and time and be conducted to further increase the number of transformants generated. The growth temperature of *A. tumefaciens* can be comprehensively evaluated by testing different temperatures (15, 17, 19, 21, 23, and 25 °C) and co-cultivation time (24, 36, 48, and 60 h) to determine when the first transformants appear and the conditions they yield the desired amount.
5. After optimizing the ATMT conditions, the transformation efficiency reaches up to 300 transformants per 10^6 conidia. In our system, the optimal transformation condition was when 9 mL of the *F. oxysporum* HS2 spore suspension (10^6 spores/mL) was mixed with an equal volume of *A. tumefaciens* LBA4404 cultures ($OD_{600} \approx 0.3$), and the 200 μL mixed cultures were co-cultivated on solid IM medium supplemented with 200 μM AS at 19 °C for 48 h (Fig. 2a–d).
6. Observation of GFP expression by fluorescence microscopy reveals that GFP is expressed constitutively in the hyphae of transformants rather than wild-type *F. oxysporum* (Fig. 3).

7. Many procedures are available to extract genomic fungal DNA. As an alternative to a commercially available kit, we follow a CTAB method [19].
8. If the T-DNA is stable in the genome, transformants will be able to grow when returned to medium containing hygromycin. An example of the results shows that all the putative transformants could grow on the selective media containing hygromycin B (Fig. 5), implying that the transformants obtained by the ATMT method are mitotically stable.

Acknowledgments

We thank Dr. Junxiang Zhang from the Chinese Academy of Agricultural Sciences for kindly providing the *Agrobacterium tumefaciens* strain LBA4404 and the pCamhybgfp vector. This work was supported partly by the National Key R&D Program of China (2016YFD0201100), the China Agriculture Research System (Cars-27), and the Hebei Natural Science Foundation (c2016204140).

References

1. Liu T, Liu L, Jiang X, Hou J, Fu K, Zhou F, Chen J (2010) *Agrobacterium-mediated* transformation as a useful tool for the molecular genetic study of the phytopathogen *Curvularia lunata*. Eur J Plant Pathol 126:363–371
2. Akashi H, Matsumoto S, Taira K (2005) Gene discovery by ribozyme and siRNA libraries. Nat Rev Mol Cell Bio 6:413–422
3. Jiang D, Zhu W, Wang Y, Sun C, Zhang KQ, Yang J (2013) Molecular tools for functional genomics in filamentous fungi: recent advances and new strategies. Biotechnol Adv 31:1562–1574
4. Lu S, Lyngholm L, Yang G, Bronson C, Yoder OC, Turgeon BG (1994) Tagged mutations at the tox1 locus of cochliobolus heterostrophus by restriction enzyme-mediated integration. Proc Natl Acad Sci U S A 91:12649–12653
5. Abuodeh RO, Orbach MJ, Mandel MA, Das A, Galgiani JN (2000) Genetic transformation of *Coccidioides* immitis facilitated by *Agrobacterium tumefaciens*. J Infect Dis 181:2106–2110
6. Rogers CW, Challen MP, Green JR, Whipps JM (2004) Use of REMI and *Agrobacterium*-mediated transformation to identify pathogenicity mutants of the biocontrol fungus, *Coniothyrium minitans*. Fems Microbiol Lett 241:207–214
7. Van SL, Denman S, Cook NC (2009) Characterisation of apple replant disease under South African conditions and potential biological management strategies. Sci Hortic 119:153–162
8. Betts MF, Tucker SL, Galadima N, Meng Y, Patel G, Li L, Donofrio N, Floyd A, Nolin S, Brown D, Mandel MA, Mitchell TK, Xu JR, Dean RA, Farman ML, Orbach MJ (2007) Development of a high throughput transformation system for insertional mutagenesis in *Magnaporthe oryzae*. Fungal Genet Biol 44:1035–1049
9. Choi J, Park J, Jeon J, Chi MH, Goh J, Yoo SY, Park J, Jung K, Kim H, Park SY, Rho HS, Kim S, Kim BR, Han SS, Kang S, Lee YH (2007) Genome-wide analysis of T-DNA integration into the chromosomes of *Magnaporthe oryzae*. Mol Microbiol 66(2):371–382
10. Dunn CN, Wang H (1998) Agrobacterium T-DNA: a silver bullet for filamentous fungi? Nat Biotechnol 16:817–818
11. Hu Y, Dai QQ, Liu YY, Yang Z, Song N, Gao XN, Ralf TV, Kang ZS, Huang LL (2014) *Agrobacterium* tumefaciens-mediated transformation of the causative agent of Valsa canker of apple tree *Valsa mali* var *mali*. Curr Microbiol 68:769–776

12. Palmero D, Rubio-Moraga A, Galvez-Patón L, Nogueras J, Abato C, Gómez-Gómez L, Ahrazem O (2014) Pathogenicity and genetic diversity of *Fusarium oxysporum* isolates from corms of *Crocus sativus*. Ind Corp Prod 61:186–192
13. De Groot MJA, Bundock P, Hooykaas PJJ, Beijersbergen AGM (1998) *Agrobacterium tumefaciens*-mediated transformation of filamentous fungi. Nat Biotechnol 16:839–842
14. Frandsen RJN (2011) A guide to binary vectors and strategies for targeted genome modification in fungi using *Agrobacterium tumefaciens*-mediated transformation. J Microbiol Meth 87:247–262
15. Geng Z, Zhu W, Su H, Zhao Y, Zhang KQ, Yang J (2014) Recent advances in genes involved in secondary metabolite synthesis, hyphal development, energy metabolism and pathogenicity in *Fusarium graminearum* (teleomorph *Gibberella zeae*). Biotechnol Adv 32:390–402
16. Islam MN, Nizam S, Verma PK (2012) A highly efficient *Agrobacterium mediated* transformation system for chickpea wilt pathogen *Fusarium oxysporum* f. sp ciceri using DsRed-express to follow root colonisation. Microbiol Res 167:332–338
17. Mullins ED, Chen X, Romaine P, Raina R, Geiser DM, Kang S (2001) *Agrobacterium*-mediated transformation of *Fusarium oxysporum*: an efficient tool for insertional mutagenesis and gene transfer. Phytopathology 91:173–180
18. Wang M-J, Li P, Wu M, Fan YS, Gu SQ, Dong JG (2012) Constuction and evaluation of ATMT mutant library of *Setosphaeria turcica*. Sci Agric Sin 45:2384–2392
19. Hallen HE, Walting R, Adams GC (2003) Taxonomy and toxicity of *Conocybe lactea* and related species. Mycol Res 107:969–979

Chapter 7

Targeted Gene Disruption Via CRISPR/Cas9 Ribonucleoprotein Complexes in *Fusarium oxysporum*

Ambika Pokhrel, Seungyeon Seo, Qiang Wang, and Jeffrey J. Coleman

Abstract

Characterization of a gene of interest frequently relies on generation of a mutant as a critical component. Transformation to disrupt a gene has been previously accomplished by several methods in *Fusarium oxysporum*. Here we provide a detailed method to generate a gene mutation mediated by a CRISPR/Cas9 ribonucleoprotein (RNP) complex. The Cas9 RNP cleaves the DNA at the target site, and during DNA repair integration of a dominant selectable marker is incorporated via homologous recombination generating the desired gene disruption.

Key words Cas9, CRISPR, Homologous recombination, Mutation

1 Introduction

Genetic transformation techniques in fungi are important tools to modify and manipulate the desired fungal isolate, thereby enabling studies to reveal the function of genes of interest and/or by inserting new genetic material into the genome of a fungus. Different methods to introduce genetic material into the fungal cells during transformation have been adapted for *Fusarium oxysporum* including PEG-mediated transformation [1, 2], *Agrobacterium*-mediated transformation [3], and electroporation-mediated transformation [4]. Recently, the use of CRISPR (clustered regularly interspaced short palindromic repeats)/Cas9-mediated genome editing coupled with PEG-mediated transformation has been developed in *F. oxysporum* [5]. Due to the ease of use, efficient gene editing, and low off-target effects, this tool has been widely implemented in many model and non-model fungi [6, 7].

CRISPR/Cas9-mediated genome editing requires two main components: a single guide RNA (sgRNA) and the Cas9

All authors contributed to this publication.

Jeffrey Coleman (ed.), *Fusarium wilt: Methods and Protocols*,
Methods in Molecular Biology, vol. 2391, https://doi.org/10.1007/978-1-0716-1795-3_7,

endonuclease. The target DNA sequence for CRISPR/Cas9 system consists of the 20 nucleotides upstream of a protospacer-adjacent motif (PAM sequence), typically an NGG sequence. The sgRNA contains the N-terminal targeting sequence which provides specificity and interacts with the Cas9 protein to form a stable complex. When a sgRNA-Cas9 complex recognizes the target DNA region, a double-stranded break is introduced between the third and fourth base upstream of the PAM and consequently activates the host DNA repair mechanisms. Three different DNA repair mechanisms can be exploited to generate mutations of the targeted gene(s): (a) non-homologous end joining (NHEJ) where error-prone repair results in random insertion or deletion of nucleotides at the cleavage site; (b) homology-dependent recombination (HDR), which involves repairing DNA based on the template homologous sequence; and (c) microhomology-mediated end joining (MMEJ), which involves DNA repair based on short homologous sequence (~50 bp) on either side of a selectable marker. In this chapter we detail the use of an in vitro assembled Cas9 ribonucleoprotein (RNP) complex coupled with a MMEJ and HDR repair template for targeted gene disruption in *F. oxysporum*.

2 Materials

2.1 Cas9 Protein Extraction and Purification

1. pHis-parallel1-NLS$_{H2B}$Cas9 (Addgene Catalog #112065) plasmid (*see* **Note 1**).
2. *E. coli* Rosetta™ (DE3) competent cells.
3. LB liquid medium: Add 10 g tryptone, 5 g yeast extract, and 5 g NaCl to 1 L water. Autoclave and store at room temperature. Aliquot 5–10 mL LB medium in two cell culture tubes.
4. 100 mg/mL ampicillin stock: Dissolve 1 g of ampicillin in 10 mL water, and sterilize through a 0.22 μm syringe filter. Store at −30 °C.
5. 100 mM IPTG stock: Add 0.238 g of IPTG to 10 mL molecular grade water. Sterilize through a syringe filter (0.22 μm), and store at −20 °C. A final concentration of 0.3 mM used.
6. Two SDS-PAGE gels and a protein electrophoresis chamber (*see* **Note 2**).
7. Ultrasonic homogenizer/sonic dismembrator (*see* **Note 3**).
8. HisPur™ Ni-NTA Chromatography Cartridges (Thermo Scientific) for purification of recombinant His-tagged proteins.
9. Buffers: Make buffers up to 500 mL with water and sterilize through a 0.45 μm filter.
   (a) Lysis buffer: 20 mM Tris–HCl, 150 mM NaCl, 10 mM imidazole, and 1 mM PMSF, pH 8.0.

(b) Wash buffer: 20 mM Tris–HCl, 500 mM NaCl, and 25 mM imidazole, pH 8.0.

(c) Elution buffer: 20 mM Tris–HCl, 500 mM NaCl, and 500 mM imidazole, pH 8.0.

10. Dialysis buffer (optional, *see* **Note 4**): 20 mM Tris–HCl and 500 mM NaCl, pH 8.0.
11. Centrifugal PES filter for 20 mL sample volume and 100 kDa molecular weight cut-off (*see* **Note 5**).
12. Water bath.
13. 37 °C shaking incubator.

2.2 sgRNA Design and Synthesis

1. CCTop web server tool (https://cctop.cos.uni-heidelberg.de/) for prediction of the CRISPR/Cas9 target site [8] (*see* **Note 6**).
2. Commercially produced sgRNA.

2.3 In Vitro Cleavage Assay

1. DNA template: PCR fragment ranging from 1 to 3 kb in size that contains the targeted cleavage site.
2. Commercially available PCR purification kit (*see* **Note 7**).
3. Synthesized sgRNA: Measure a concentration and store at −80 °C.
4. Purified Cas9 protein: Measure a protein concentration and store at −30 °C.
5. Cas9 nuclease buffer: 20 mM HEPES, 100 mM NaCl, 5 mM $MgCl_2$, and 0.1 mM EDTA, pH 6.5 (*see* **Note 8**).
6. DEPC water (0.1% DEPC in 100 ml water) or nuclease-free water.
7. 200 μL PCR tubes.
8. 0.8% agarose gel.

2.4 Preparation of Donor DNA

2.4.1 ~1 kb Overlapping Flanking Region for Homologous Recombination Strategy

1. pUC19 vector.
2. Restriction enzymes: BamHI and HindIII (*see* **Note 9**).
3. PCR purification kit (*see* **Note 7**).
4. Genomic DNA of *F. oxysporum* for use as a PCR template.
5. pCWHyg1 plasmid (*see* **Note 10**).
6. Primers to generate three PCR products (a ~1 kb upstream fragment, a ~1 kb downstream fragment, and the Hyg^R cassette fragment) for use in plasmid assembly (*see* **Note 11**).
7. Gel extraction or PCR purification kit.
8. NEBuilder HiFi DNA assembly kit (New England Biolabs, Ipswich, MA) to assemble the plasmid.
9. Competent DH5α *E. coli* cells.

10. LB agar plates.
11. 100 mg/mL ampicillin.
12. 100 mM IPTG. A final concentration 1 mM IPTG is used.
13. 20 mg/mL X-Gal: Dissolve 200 mg of X-Gal (5-bromo-4-chloro-3-indolyl-β-D-galactoside) in 10 mL dimethylsulfoxide. Mix well and aliquot to microcentrifuge tubes. Cover the tubes with aluminum foil and store at −20 °C. The final concentration of X-Gal used is 200 μg/mL in the media.
14. Commercially available plasmid extraction kit.
15. Thermocycler.
16. 37 °C incubator.

2.4.2 50 bp Overlapping Region Strategy

To prepare the 50 bp templates for transformation, the pCWHygl plasmid (selectable marker), commercial PCR purification kit, and thermocycler are required as mentioned in the 1 kb strategy (Subheading 2.4.1).

1. 50-bp overhang primers.
   (a) Forward primer: Combine a 50 bp upstream homologous sequence before the Cas9 RNP cut site with the N-terminus of the hygromycin cassette (containing the *glaA* promoter).
   (b) Reverse primer: Add a 50 bp downstream homologous sequence after the Cas9 RNP cleavage site to the C-terminus of the hygromycin cassette (containing the terminator), and make reverse complementary sequences.

2.5 Growth of *Fusarium oxysporum*

1. *Fusarium oxysporum* isolate of interest (*F. oxysporum* isolates with sequenced genomes can be obtained from the Fungal Genetics Stock Center [9]).
2. Potato dextrose broth (PDB) medium: Add 2.4 g PDB to 100 mL deionized water, and autoclave in a 250 mL Erlenmeyer flask. Keep at room temperature.
3. Potato dextrose agar (PDA) plates: Add 24 g PDB and 22 g agar to 1 L deionized water and autoclave. Pour the medium into 10 cm plates (*see* **Note 12**).
4. A platform shaker at 160 rpm at room temperature.
5. Hemocytometer and compound microscope.

2.6 Making Protoplasts

1. Potato dextrose broth (PDB) medium: Same as above.
2. A sterile nylon mesh (37 μm pore size) in a funnel.
3. Whatman filter paper grade 4 circles and 55 mm diameter.
4. Buchner funnel with a sidearm flask and tubes for vacuum filtration.

5. A sterile round spatula.
6. 6 mg/mL kitalase solution in 0.7 M NaCl (*see* **Note 13**).
7. Rotary shaker.
8. Hemocytometer and compound microscope.
9. SuTC buffer: 20% sucrose, 50 mM Tris–HCl, pH 7, and 50 mM $CaCl_2$. Autoclave and store at 4 °C.

2.7 Fusarium Transformation

1. PSTC buffer: 60% polyethylene glycol (PEG, molecular weight 3350), 20% sucrose, 50 mM Tris–HCl, pH 7, and 50 mM $CaCl_2$. Mix all components in a 50 mL Falcon tube, and place in a 65 °C water bath to dissolve (*see* **Note 14**). Sterilize through a 0.22 μm syringe filter.
2. TB3 medium: 20% sucrose, 1% glucose, 0.3% yeast extract, and 0.3% casamino acids. Autoclave and store at 4 °C.
3. Regeneration top (RTop) agar plate: 20% sucrose, 1% glucose (dextrose), 1% yeast extract, 1% casamino acids, and 1.5% agarose. Prepare 100 mL of the RTop. Autoclave and keep in a 55–65 °C water bath until use.
4. PDA bottom agar plate: 2.4 g PDB, 2.2 g agar up to 100 mL of water. Cool the autoclaved media and add hygromycin B prior to usage.
5. 100 mg/ml hygromycin B stock: Aliquot to the microcentrifuge tube and store in 4 °C. A final concentration of 120 μg/ml used (*see* **Note 15**).
6. 55–65 °C water bath.

3 Methods

3.1 Cas9 Purification

1. Transform the plasmid (pHis-parallel1-NLS_{H2B}Cas9) into the *E. coli* Rosetta™ (DE3) strain for high expression (*see* **Note 16**) (Fig. 1).
2. Pick a single colony and inoculate to 10 mL LB liquid medium containing 100 μg/mL ampicillin, and allow to grow for 14 h in a 37 °C shaking incubator.
3. Transfer the cultured medium to 1 L of fresh LB liquid medium, and continue to incubate the culture until the OD_{600} value reaches 0.6. Add IPTG to the medium to a final concentration of 0.3 mM, and incubate overnight at 25 °C, 160 rpm on a rotary shaker.
4. Check the Cas9 protein expression through SDS-PAGE gel (*see* **Note 2**).
5. Pellet the bacteria by centrifugation at 2000 g for 15 min, and decent the supernatant.

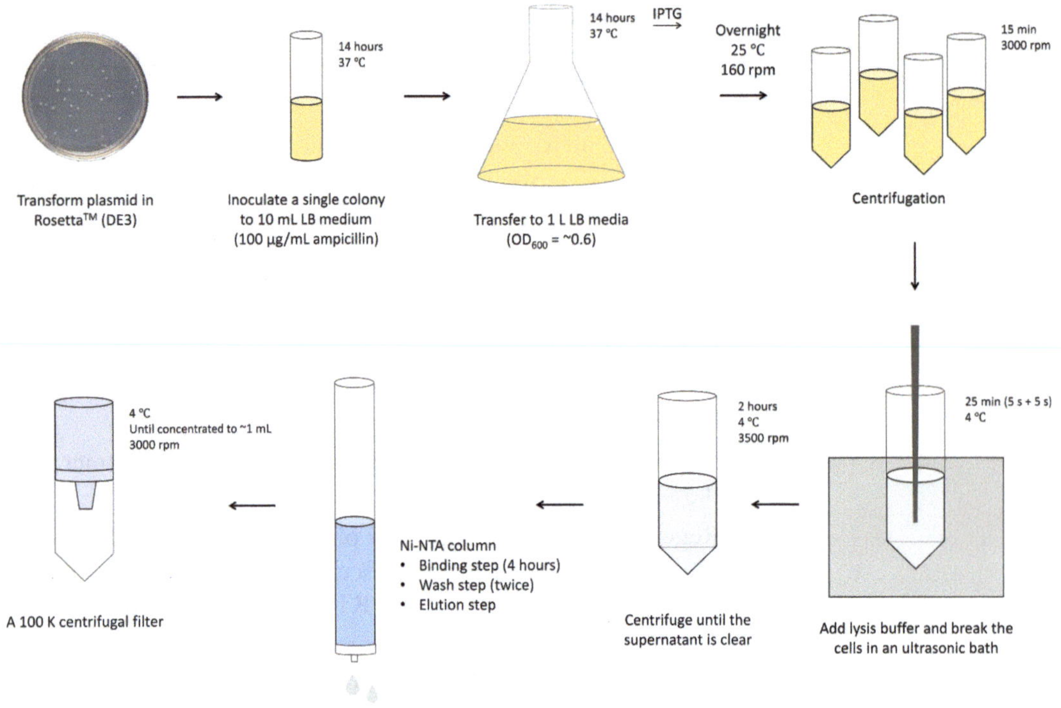

Fig. 1 The schematic representation of Cas9 protein purification method using a HisPur Ni-NTA column

6. Add 30 mL lysis buffer and suspend the pellet. Proceed to lyse the *E. coli* cells in an ultrasonic bath (5 s + 5 s, 25 min) (*see* **Note 3**).
7. Centrifuge the mixture at 2500 × *g*, 4 °C for 2 h until the supernatant is clear. Filter the supernatant with a 0.22 μm filter syringe.
8. Transfer the filtered mixture to the Ni-NTA column, and enable the His tagged Cas9 protein to bind to the Ni column by incubating for 4 h on a rotary shaker at 4 °C (*see* **Note 17**).
9. Wash the column twice with 10–15 mL (~2× resin volume) of the wash buffer.
10. Elute the Cas9 protein with the addition of 20–50 mL elution buffer (*see* **Note 18**).
11. Concentrate the Cas9 protein to a final concentration to 4 mg/mL by adding the eluted protein to a 100 K centrifugal filter, and centrifuge at 3000 rpm at 4 °C until protein is concentrated to ~1 mL (*see* **Note 5**).
12. Check the activity of Cas9 protein by in vitro cleavage (*see* Subheading 3.3).
13. Store the final protein at −80 °C until use.

3.2 sgRNA Design and Synthesis

1. Design a target-specific oligo(s) for the gene disruption using CCTop (*see* **Note 6**) (Fig. 2a). Take into considerations the following parameters:
   (a) A sgRNA target site should be located close to the 5′ end of the gene coding sequence.
   (b) To avoid non-specific cleavage of other loci (off-target), at least 3 potential target sequences with a high efficiency score should be selected and searched by BLASTN against the *F. oxysporum* genome.

3.3 In Vitro Cleavage Assay

1. Generate a 1 to 3 kb PCR product containing the sgRNA target site. Check the PCR fragment by 1% agarose gel, and purify the PCR fragment using a commercially available PCR purification kit (*see* **Note 7**).
2. Thaw the prepared sgRNA, Cas9 protein, DNA template, buffers, and RNase-free water on ice. Calculate the necessary amount (μg) of each component with the ratio 1 sgRNA: 5 Cas9 protein: 0.5 DNA template.
3. Add the 10× Cas9 nuclease buffer (2 μL), 200 ng sgRNA, 1 μg Cas9, 100 ng of DNA template, and DEPC water to a final volume of 20 μL. Mix thoroughly by pipetting.
4. Incubate at 37 °C for 1 h.
5. Visualize the cleavage activity by gel electrophoresis on a 0.8–1% agarose gel (*see* **Note 19**).

3.4 Preparation of Donor DNA

3.4.1 1000 bp Overlapping Region Strategy

1. Amplify 1 to 3 kb upstream and downstream homologous regions from *F. oxysporum* genomic DNA (Fig. 2b). Generate a PCR product containing the *gpdA* promoter, hygromycin phosphatase gene, and terminator from pCWHyg1. Purify PCR products using the PCR purification kit (*see* **Note 7**).
2. Digest pUC19 using BamHI and HindIII restriction enzymes following the manufacturer's instruction (*see* **Note 9**).
3. Assemble the linear pUC19 and three PCR fragments (the upstream flanking sequence, the hygromycin resistance cassette, and the downstream flanking sequence) using the NEBuilder HiFi DNA assembly kit (*see* **Note 20**).
4. Incubate at 50 °C for 1 h.
5. Transform 5 μL of the assembled product, and mix with competent DH5α *E. coli* cells (*see* **Note 16**).
6. Extract plasmid using a commercially available kit following the manufacturer protocol (*see* **Note 21**).
7. Confirm the desired fragments inserted into the vector by PCR and DNA sequencing.
8. Measure the concentration of the construct and store at −30 °C.

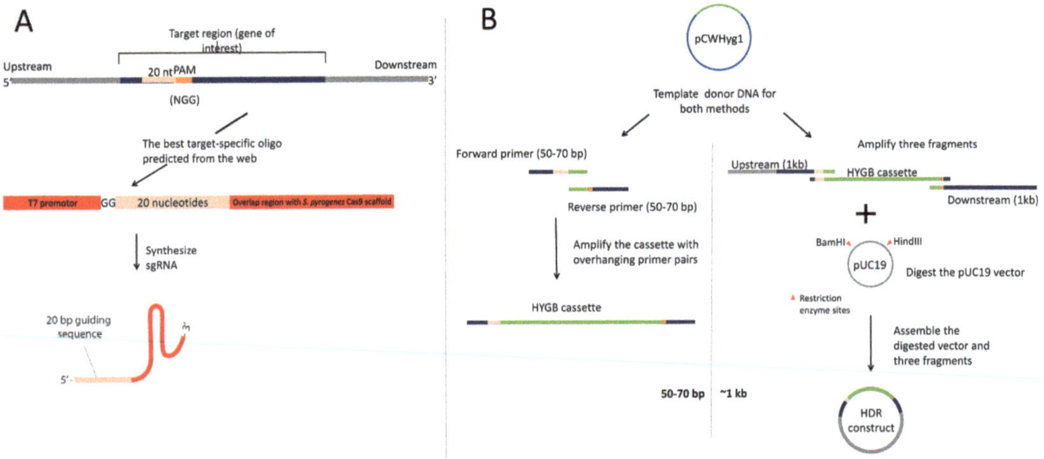

Fig. 2 Schematic diagram of sgRNA and donor DNA construction. (**a**) sgRNA target site with PAM sequence is selected using the CCTop server. To the 20 nucleotide target sequence (without NGG), T7 promoter is added to the 5′ end, and a 14 nucleotide overlapping sequence with *S. pyogenes* is added to the 3′ end. If the 5′ end of the 20 nucleotide sequence lacks G's, two G's are added. The sgRNA is synthesized using the commercially available kit, and the cleavage efficiency is checked. (**b**) Donor DNA is constructed in two different ways using the pCWHyg1 plasmid. For the 50–70 bp strategy, the hygromycin cassette is amplified using primer pairs having the 50–70 bp overhanging regions. For the 1 kb strategy, 1 kb upstream, 1 kb downstream, and the hygromycin cassette are amplified using different sets of overlapping primers. These three fragments are assembled into the digested pUC19 vector (digested with restriction enzymes BamHI and HindIII). This construct is confirmed by PCR screening and sequencing with M13F and M13R primers

3.4.2 50–70 bp Overlapping Region Strategy

1. Amplify the hygromycin cassette with primer pairs having a 50–70 bp overhanging region (Fig. 2b). Purify PCR products using a commercially available PCR purification kit.
2. Measure the concentration of the PCR product and store at −30 °C until use.

3.5 Protoplasting

1. Culture the *F. oxysporum* isolate in 100 mL of PDB media for 5 days at room temperature on a platform shaker at 160 rpm.
2. Remove the hyphae by filtering the culture through a sterile nylon mesh, and collect the conidia that flowed through the mesh by centrifugation at 2000 × *g* for 10 min. Count the conidia using a hemocytometer.
3. Inoculate 10^8 conidia into fresh PDB media, and let it grow in the shaker at 160 rpm overnight (12–15 h).
4. Monitor conidia germination by microscopy. Proceed to next step if most have germinated (*see* **Note 22**).
5. Collect the germinated conidia on a sterile Whatman filter paper by vacuum filtration (*see* **Note 23**). Transfer the germlings to a 50 mL centrifuge tube using a sterile spatula.
6. Weight the germinated conidia and suspend 1 g in 15 mL of 0.7 M NaCl containing 6 mg/mL kitalase (*see* **Note 24**).

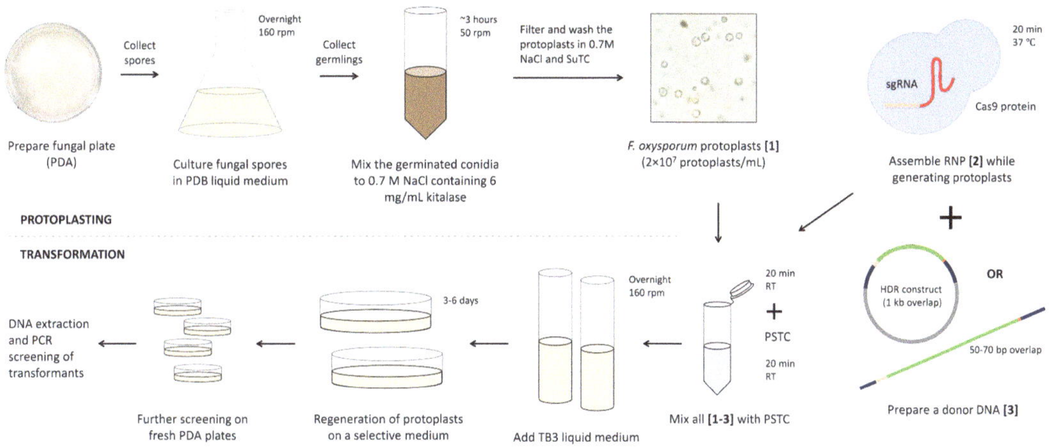

Fig. 3 Schematic representation of PEG-mediated delivery of Cas9 RNPs to *Fusarium oxysporum* protoplasts in order to generate the specific targeted gene disruption

7. Incubate the mycelium-enzyme suspension at room temperature for 2 h on a rotary shaker at 50 rpm. Pipet a 10 μL sample of the protoplast reaction to a hemocytometer, and check the protoplast quality and quantity in the microscope. If necessary, continue to incubate longer and check every 30 min (Fig. 3).
8. Filter the protoplasts through the sterile nylon mesh, and collect by centrifugation at 1000 × *g* at 4 °C for 10 min (*see* **Note 25**).
9. Wash the protoplasts once by suspending the protoplasts in 10 mL of 0.7 M NaCl, and centrifuge at 1000 × *g* at 4 °C for 10 min.
10. Suspend the protoplasts with 10 mL of SuTC buffer, and collect the protoplasts by centrifugation at 1000 × *g* at 4 °C for 10 min.
11. Suspend the protoplasts in 1 mL SuTC, and count the number of protoplasts by hemocytometer. Dilute protoplasts in SuTC to a final concentration of 2×10^7 protoplasts/mL.

3.6 Transformation

1. Prepare the donor DNA and Cas9 RNPs while the protoplasts are being generated.
2. In a 0.2 mL PCR tube, make the Cas9 RNPs with a total reaction volume of 50 μL (1× Cas9 nuclease buffer, 40 μg Cas9, 8 μg sgRNA, and DEPC water). Mix well and incubate at 37 °C for 20 min.
3. Mix 200 μL of the protoplast mixture with 50 μL of the Cas9 RNPs and 6 μg of the donor DNA, and incubate at room temperature for 20 min.
4. Add 1 mL of PSTC to the protoplast mixture, and mix gently by pipetting and incubate for 20 min at room temperature.

5. Add 3 mL of TB3 and allow the protoplasts to regenerate by shaking at 160 rpm for 18 to 20 h.
6. Centrifuge the regenerated protoplasts at 2000 × *g* for 10 min, and suspend in 100 μL of SuTC.
7. Mix the SuTC protoplast suspension with 10 mL RTop medium (at 55 °C), and pour into a 10 cm petri plate containing PDA with 120 μg/ml hygromycin B. Incubate the transformation plates in the dark at room temperature.
8. After 3~6 days, hygromycin-resistant transformants will begin to appear on the agar surface. Transfer individual single colonies (putative hygromycin-resistant transformants) to fresh PDA plates containing hygromycin for further screening for the desired mutation.

4 Notes

1. This expression plasmid has a *Streptococcus pyogenes* Cas9 protein fused with the nuclear localization signal (NLS) from histone H2B from *F. oxysporum*. Since this NLS sequence is conserved in *Fusarium* spp., this Cas9 should function well within most *Fusarium* isolates.
2. A common SDS-PAGE gel is recommended to be used to monitor the induction and expression of the Cas9 gene at various time points throughout the purification. The molecular weight of the recombinant protein is approximately 165 kDa. Thus, a ~6% gel was used in this study.
3. An ultrasonic homogenizer is recommended to lyse the bacterial cells, although multiple other methods can be utilized including lysozyme, detergents, etc. To avoid the Cas9 protein from being degraded during purification, the lysis buffer containing the protein should be kept on ice or at 4 °C during lysis.
4. To remove the imidazole, which is a key component during protein purification but also harbors antifungal activity, use a dialysis bag and the same composition of buffer without the imidazole. Conduct the dialysis on ice or at 4 °C to prevent protein degradation.
5. Based on the volume of the extracted protein, choose the right size of centrifugal filter. It is recommended to choose a filter with a molecular weight cut-off of 100 kDa to aid in further purification of the Cas9 protein. The centrifugation step to concentrate the sample will take a various length of time based on the quantity/quality of protein.
6. Other web-based CRISPR/Cas9 target sequence tools are available including CHOPCHOP [10] (https://chopchop.

cbu.uib.no/) and CRISPick (https://portals.broadinstitute.org/gpp/public/analysis-tools/sgrna-design).

7. The use of gel extraction for PCR product purifications is highly recommended, as unintended PCR products can be present even if not visualized after gel electrophoresis. The products affect further assays such as plasmid assembly, in vitro cleavage, and eventual transformation. Most commercial kits have performed well.
8. The Cas9 nuclease buffer can be made as described, although commercially available buffers such as the NEBuffer 3.1 and the Cas9 nuclease reaction buffer from New England Biolabs are also suitable.
9. Other restriction enzymes can be chosen, if they perform well in the multiple cloning site (MCS) region of pUC19.
10. Hygromycin phosphotransferase is most commonly used as the selectable marker for *F. oxysporum*, but other selectable markers can be used such as geneticin (G 418). Other plasmids containing a hygromycin resistance cassette are suitable.
11. Overlapping regions $\geq$20 bp are required to avoid improper assembly. Usually 25 nucleotides as the minimal overlap are sufficient.
12. Most fungal strains grow well on PDA. However, culturing on different media may be necessary to generate the micro/macroconidia required for germinating overnight.
13. Kitalase (Wako Chemicals, Japan) is a commercial enzyme for protoplast and spheroplast formation in fungi. It can be prepared by dissolving in other solutions including 1.2 M KCl, 1.2 M sorbitol, as well as 0.7 M NaCl based on fungal strains. Germlings incubated in the presence of 6 mg/mL of kitalase usually produce protoplasts within 3 h, although production of protoplasts should be monitored.
14. When polyethylene glycol (PEG) dissolves in water, the volume increases. Thus, add only ¼ the desired volume of water initially. The PEG solution can be warmed in a 65 °C water bath to aid in dissolving the PEG solution. After the PEG is completely dissolved, more water can be added to the desired final volume.
15. The concentration of the selective compound (hygromycin, geneticin, etc.) used for selecting putative transformants should be evaluated by testing growth of the wild type isolate under different concentrations of the compound.
16. The procedure for the bacterial transformation is explained briefly: Add 3 μL of DNA (i.e., pHis-parallel1-NLS_{H2B}Cas9) into 50 μL of *E. coli* competent cells in a microcentrifuge tube. Mix gently by flicking the bottom of the tube, and incubate the mixture of DNA and competent cells on ice for 20 min. Heat

shock the tube by placing the tube into a 42 °C water bath for 45 s, and return to ice for 1 min. Add 1 mL LB media and grow in a 37 °C shaking incubator for 1 h. Centrifuge for 1 min to pellet the bacterial cells and discard the supernatant. Suspend the cells in 200 μL of fresh LB, and plate them on a LB agar plate with ampicillin (100 μg/mL). Incubate plates at 37 °C incubator until colonies appear.

17. The Cas9 protein in this study is a 6× His-tagged recombinant protein. Thus, the recombinant protein is captured by the Ni-NTA matrix; however other purification systems for 6× His-tagged proteins are suitable.
18. To prevent losing Cas9 protein, protein detection assays such as the Bradford can be used to ensure that all the Cas9 protein was completely eluted.
19. Depending upon multiple factors including the quality of purification of the Cas9 protein, fungal isolate, and/or Cas9/ sgRNA efficiency, the Cas9 protein: sgRNA ratios can differ. A step to find the proper ratio that works best is recommended and the same ratio should be used for further transformation.
20. Other DNA assembly kits/systems can be used.
21. Any other plasmid extraction kit for *E. coli* can be used.
22. Germlings that are 3–4 times the size of conidia are ideal for enzyme treatment.
23. A sterile nylon mesh (~37 μm pore size) can also be used to collect the germlings instead of the Whatman filter paper.
24. Collected germinated conidia weighing 0.5–1 g are usually enough to prepare the necessary amount of protoplasts. The prepared kitalase solution can be stored at −20 °C and used multiple times with the same *F. oxysporum* isolate. The time required to generate protoplasts will increase due to decreased enzymatic activity of the kitalase.
25. After the kitalase treatment, protoplasts must be handled gently.

Acknowledgments

Development of this method and publication was partially supported by Cotton Inc., the Alabama Agricultural Experiment Station, and the Hatch program of the National Institute of Food and Agriculture, USDA.

References

1. Kistler HC, Benny UK (1988) Genetic transformation of the fungal plant wilt pathogen, *Fusarium oxysporum*. Curr Genet 13:145–149
2. Coleman JJ, Wasmann CC, Usami T, White GJ, Temporini ED, McCluskey K, VanEtten HD (2011) Characterization of the gene encoding pisatin demethylase (*FoPDA1*) in *Fusarium oxysporum*. Mol Plant-Microbe Interact 24:1482–1491
3. Mullins ED, Chen X, Romaine P, Raina R, Geiser DM, Kang S (2001) Agrobacterium-mediated transformation of *Fusarium oxysporum*: an efficient tool for insertional mutagenesis and gene transfer. Phytopathology 91:173–180
4. Liang L, Li J, Cheng L, Ling J, Luo Z, Bai M, Xie B (2014) A high efficiency gene disruption strategy using a positive–negative split selection marker and electroporation for *Fusarium oxysporum*. Microbiol Res 169(11):835–843
5. Wang Q, Cobine PA, Coleman JJ (2018) Efficient genome editing in *Fusarium oxysporum* based on CRISPR/Cas9 ribonucleoprotein compleses. Fungal Genet Biol 117:21–29
6. Wang Q, Coleman JJ (2019) Progress and challenges: development and implementation of CRISPR/Cas9 technology in filamentous fungi. Comput Struct Biotechnol J 17:761–769
7. Schuster M, Kahmann R (2019) CRISPR-Cas9 genome editing approaches in filamentous fungi and oomycetes. Fungal Genet Biol 130:43–53
8. Stemmer M, Thumberger T, del Sol Keyer M, Wittbrodt J, Mateo JL (2015) CCTop: an intuitive, flexible and reliable CRISPR/Cas9 target prediction tool. PLoS One 10:e0124633
9. McCluskey K, Wiest A, Plamann M (2010) The Fungal Genetics Stock Center: a repository for 50 years of fungal genetics research. J Biosci 35(1):119–126
10. Labun K, Montague TG, Krause M, Torres Cleuren YN, Tjeldnes H, Valen E (2019) CHOPCHOP v3: expanding the CRISPR web toolbox beyond genome editing. Nucleic Acids Res 47(W1):W171–W174. https://doi.org/10.1093/nar/gkz365

Chapter 8

CRISPR/Cas9 RNP-Mediated Gene Fusion to Assess Protein Quantification and Subcellular Localization in *Fusarium oxysporum*

Qiang Wang and Jeffrey J. Coleman

Abstract

F. oxysporum is a notorious filamentous pathogenic fungus that causes serious problems in agriculture and animal/human health. Knowing how the fungus interacts throughout the course of an infection is necessary to propose an effective control strategy, and consequently the manipulation of the *F. oxysporum* genome is essential to investigate the molecular interplay between the host and fungus. To facilitate assessing protein quantification and subcellular localization, we developed a simple, economical CRISPR/Cas9-mediated endogenous gene tagging (EGT) system based on two different strategies, homology-independent targeted integration (HITI) and homology-dependent recombination integration (HDRI). Reporter genes, including *GFP* and *LacZ*, can be inserted at the N- or C-terminus of an endogenous gene of interest at the original chromosomal locus, allowing partial characterization of the gene function.

Key words Gene fusion, Tagging, GFP, mCherry, Localization

1 Introduction

Fusarium oxysporum is an important soil-borne fungal pathogen responsible for infections on a wide range of host plants and resulting in the agriculturally significant *Fusarium* wilt disease [1]. Studies of the molecular interactions between *F. oxysporum* and the host plants are critical for understanding fungal pathogenicity and proposing an effective agricultural control strategy. Unlike many módel fungi, the molecular manipulation of *F. oxysporum*, including gene disruption/deletion and generation of gene fusions or tagging of exogenous genes for protein subcellular localization and expression can often be difficult to conduct.

CRISPR/Cas9 genome editing technology is becoming applicable to filamentous fungi and largely enhances the success rates of genome manipulation. In *F. oxysporum*, a CRIPSR/Cas9

Jeffrey Coleman (ed.), *Fusarium wilt: Methods and Protocols*,
Methods in Molecular Biology, vol. 2391, https://doi.org/10.1007/978-1-0716-1795-3_8,

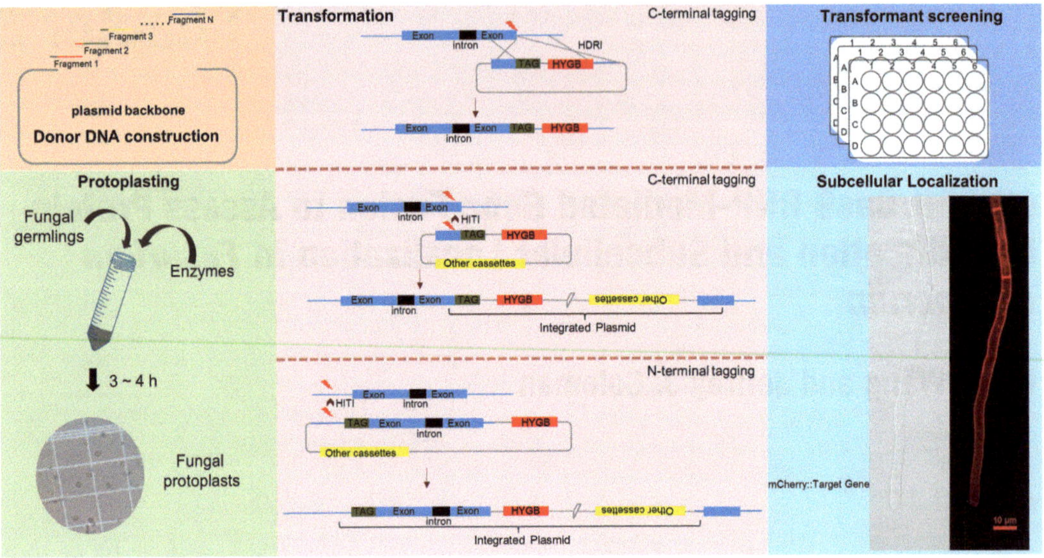

Fig. 1 The pipeline of CRISPR/Cas9 RNP-mediated gene tagging to assess protein quantification and subcellular localization in *F. oxysporum*

ribonucleoprotein protocol has been established to conduct gene deletions/disruptions [2]. One obvious benefit with the use of CRISPR/Cas9 technology is that it allows precise genome modifications, enabling endogenous gene tagging (EGT). EGT can be used to fuse a fluorescent or reporter gene (e.g., *GFP*, *mCherry*, or *lacZ*) to the N- or C-terminus of an endogenous open reading frame for investigating the subcellular localization and/or expression dynamics of the fusion protein(s) under the endogenous regulatory control [3, 4]. In this chapter, a detailed protocol to conduct endogenous gene tagging in *F. oxysporum* is provided using two different strategies, homology-dependent recombination integration (HDRI) and homology-independent targeted integration (HITI), based on the CRISPR/Cas9 RNP transformation method (Fig. 1).

2 Materials

2.1 Fungal Isolates and Bacterial Strains

1. Fungal strain: *Fusarium oxysporum* isolate to be studied. Several isolates with publically available genome sequences are available through the Fungal Genetics Stock Center [5].
2. Bacterial strain: DH5α or Top10 *Escherichia coli* strain used for construction of donor DNA.

2.2 Fungal and Bacterial Culture Media

1. Potato Dextrose Broth medium (PDB): Add 24 g potato dextrose broth to 1 L deionized water and autoclave.
2. Potato Dextrose Agar medium (PDA): Add 24 g potato dextrose broth and 20 g agar to 1 L deionized water and autoclave. Pour the medium into 10 cm culture plates.
3. Luria-Bertani (LB) medium: For liquid medium, add 10 g peptone, 5 g yeast extract, and 5 g NaCl to 1 L deionized water and autoclave. For solid medium, add 15 g agar with above reagents to 1 L deionized water and autoclave.
4. YG medium: Add 10 g yeast extract and 30 g glucose to 1 L deionized water and autoclave.
5. M-100 trace element solution: Dissolve 30 mg H_3BO_3, 70 mg $MnCl_2 \cdot 4\ H_2O$, 200 mg $ZnCl_2$, 20 mg $NaMoO_4 \cdot 2\ H_2O$, 50 mg $FeCl_3 \cdot 6\ H_2O$, 200 mg $CuSO_4 \cdot 5\ H_2O$ in 500 mL H_2O. Autoclave, protect from light, and store at 4 °C.
6. M-100 salt solution: Mix 16 g KH_2PO_4, 4 g Na_2SO_4, 8 g KCl, 2 g $MgSO_4 \cdot 7\ H_2O$, 1 g $CaCl_2$ (*see* **Note 1**), 8 mL trace element solution, and add H_2O to 1 L. Autoclave and store at 4 °C.
7. M-100 medium (minimum medium) [6]: 10 g glucose, 3 g KNO_3, 62.5 mL M-100 salt solution, 937.5 mL H_2O, and 20 g agar. Autoclave and pour to 10 cm plates.
8. Bottom agar: Add 3.0 g yeast extract, 3.0 g casamino acids, 200 g sucrose, and 10 g agar to 1 L deionized water, and autoclave.
9. Top agar: Add 3.0 g yeast extract, 3.0 g casamino acids, 200 g sucrose, and 15 g agar to 1 L deionized water, and autoclave.
10. TB3 medium: Add 3.0 g yeast extract, 3.0 g casamino acids 3.0 g, and 200 g sucrose to 1 L deionized water, and autoclave.
11. STC buffer: Add 200 g sucrose, 7.35 g $CaCl_2 \cdot 2\ H_2O$, and 100 mL 0.5 M Tris–HCl (pH 7.0) to 1 L deionized water, and autoclave.

2.3 Donor DNA Construction

1. New England Biolabs NEBuilder® HiFi DNA Assembly Kit.
2. High Fidelity PCR master mix and primers.
3. Commercial gel extraction kit and/or PCR purification kit.
4. LB plates containing X-gal, IPTG, and appropriate selection antibiotic.
5. Commercial plasmid extraction kit.
6. Agarose.
7. Gel electrophoresis chamber, casting try, well mold, and power supply.
8. Thermocycler.
9. Platform rotary shaker.

2.4 Transformation Reagents

1. 60% Polyethylene glycol (PEG): Add 6 g PEG 6000 into 4 mL STC buffer, and place in a 65 °C water bath until dissolved. The 60% PEG solution is filtered sterilized using a 0.22 μM filter.
2. 1.2 M KCl buffer: Add 89.46 g KCl to 1 L deionized water and autoclave.
3. Hygromycin stock: 100 mg/mL. Store at 4 °C.
4. Ampicillin stock: 100 mg/mL. Store at −20 °C.
5. Protoplasting enzyme solution: Add 0.1 g Kitalase (final concentration: 5 mg/mL, Wako Pure Chemical Industries, Ltd.) into 20 mL 1.2 M KCl buffer, and dissolve at room temperature for 20 min. Filter sterilize the protoplasting enzymatic solution with a 0.22 μM syringe filter.
6. Commercially available sgRNA: The sgRNA sequence can be obtained using CCTop (https://cctop.cos.uni-heidelberg.de/) [7] which predicts CRISPR/Cas9 cleavage sites within the input sequence.
7. Cas9 protein: Cas9 protein purified using HisPur™ Ni-NTA Chromatography Cartridges (Thermo Fisher Scientific) with the plasmid, pHis-parallel1-NLSH2BCas9 (Addgene#112065).
8. 10× Cas9 Nuclease Reaction Buffer.
9. Hemocytometer.
10. Sterile funnels containing a sheet of Nitex nylon mesh (40 μm).
11. Sterile 15 mL conical centrifuge tubes.
12. Sterile 50 mL conical centrifuge tubes.

2.5 Microscopy

1. 4% paraformaldehyde.
2. Phosphate buffered saline (PBS), pH 7.4.
3. 2.5% water agar Petri plates.
4. Sterile scalpel.
5. Glass slides.
6. Cover slips.
7. Confocal microscope (this method was initially developed using a Nikon Al confocal microscope).

3 Methods

3.1 Construction of Donor DNA

The construction of the donor DNA is dependent on the location of the desired fusion, either the N- or C-terminus of the protein. For generation of a C-terminus tag using the HDRI strategy, a ~ 500 bp fragment containing the locus encoding the C-terminus of the gene of interest is amplified and fused with the desired marker/tag (GFP, mCherry, etc.), a universal terminator, a hygromycin cassette, and the homologous fragment (~500 bp) after the sgRNA site. To generate a C-terminus tag using a HITI-based method, a ~ 500 bp fragment located in the coding region of the C-terminus of the protein of interest containing the sgRNA cleavage site is amplified and fused with the marker/tag, a universal terminator, and the hygromycin cassette. Generation of a N-terminus tag using a HITI strategy requires a fragment at the front of the start codon of a gene, and containing the sgRNA site was cloned and fused with a tagging marker, the gene ORF with 3′-UTR and the hygromycin cassette.

1. A common cloning plasmid (for instance, pUC19) as a background plasmid is chosen to generate the donor plasmid and is linearized by digestion with the restriction enzymes HindIII and BamHI.
2. Primer design is conducted based on the NEB online tool, NEBuilder (http://nebuilder.neb.com/#!/). The NEB assembly strategy requires at least 25 bp overlap of the ends between one fragment and the following fragment (*see* **Note 2**).
3. All fragments are amplified by PCR using a high-fidelity polymerase and gel purified using a commercially available gel extraction kit.
4. The plasmid assembly is conducted by simultaneously mixing equal molar concentrations of the PCR fragments and linearized background plasmid with 2× NEBuilder HiFi DNA Assembly Master Mix with a 50 °C incubation for 1 h.
5. The above reaction mixture (~3 μL) is used in transformation of *E. coli* (DH5α or Top10 strains are suitable), and plated on LB plates containing the X-gal and the appropriate selection antibiotic based on the selected background plasmid. Incubate the plated overnight at 37 °C for growth of transformants.
6. If selecting using X-gal and blue/white screening, isolate individual colonies to a new LB plate containing X-gal and appropriate selection antibiotic. Place plate in 37 °C incubator for overnight growth.
7. Select individual colonies for screening by inoculating 3 mL of LB medium in a culture tube, and grow overnight at 37 °C on a rotary shaker.
8. Screen colonies for desired construct by either PCR and/or restriction digest of purified plasmids.

3.2 In Vitro Cas9 RNP Cleavage Assay

1. A ~ 2 kb DNA fragment of the locus containing the Cas9 RNP cleavage site is amplified from genomic DNA by PCR. The PCR product is purified using a commercial PCR purification kit, and the concentration determined by NanoDrop (*see* **Note 3**).

2. A cleavage assay is conducted in a 20 μL total volume which is composed of 1 μL of the purified Cas9 protein (200 ng/μL), 40 ng sgRNA, 1× Cas9 Nuclease Reaction Buffer, 3 μL of the purified PCR fragment (20 nM), and nuclease-free water (*see* **Note 4**).
3. The tube containing the in vitro cleavage reaction is incubated at 37 °C for 1 h.
4. The final products are analyzed by electrophoresis in a 0.8% agarose gel. Efficient cleavage of the DNA at the target site is apparent by the presence of two DNA fragments when compared to the original PCR product on an agarose gel.

3.3 Generating Fungal Protoplasts

1. Inoculate the *F. oxysporum* fungal isolate to 100 mL liquid M-100 medium, and allow to grow for 5 days on a shaker (180 rpm) to produce conidia (*see* **Note 5**).
2. Harvest conidia by filtering through Nitex (~40 μm pore size) or through three layers of cheese cloth. Count conidia using a hemocytometer to determine concentration.
3. Inoculate 10^8 conidia into 100 mL PDB, and allow the conidia to germinate overnight on a shaker (~12 h, *see* **Note 6**).
4. Harvest the germlings using a Nitex nylon mesh (40 μm), and wash the germlings with sterile deionized water to remove the un-germinated conidia.
5. Incubate the germlings in the protoplasting enzyme solution in a 50 mL conical centrifuge tube for 3–4 h on a rotary shaker at 50 rpm (*see* **Note 7**). The ratio of the germlings to the enzymatic solution should be ~1/20 g/mL. Monitor protoplast generation every 30 min after 2 h incubating in the solution (*see* **Note 8**). Once most of the germlings are protoplasts, proceed to the next step.
6. Filter the protoplasts through a Nitex nylon mesh (40 μM), and collect the protoplasts by centrifugation at 1000 × *g* for 8 min at 4 °C. Remove the supernatant.
7. Add 10 mL STC buffer and gently suspend the protoplasts in the solution by pipetting. Centrifuge the protoplasts as before, 1000 × *g* for 8 min at 4 °C. Repeat this step one more time for a second wash with STC.
8. Suspend protoplasts in STC buffer at a final concentration of 2–5 × 10^7/ mL. Use 200 μL of the protoplast solution for each transformation (*see* **Note 9**).

3.4 PEG-Mediated Transformation

1. The Cas9 RNPs are prepared using a 1:1 molar ratio of Cas9 and sgRNA. This solution is mixed in a 30–50 μL total volume with 10× Cas9 Nuclease Reaction Buffer and DEPC-treated water (*see* **Note 4**).

2. Incubate Cas9 RNPs at room temperature for 20 min.
3. Add Cas9 RNPs and the donor DNA (1–3 μg, < 10 μL) to 200 μL of the protoplast solution, and incubate the above solution for 5–20 min at room temperature (*see* **Note 10**).
4. Add an equal volume of 60% PEG to the protoplast/Cas9 RNP solution, mix by gentle pipetting, and allow to incubate at room temperature for 20–30 min.
5. The above solution is transferred to a 15 mL conical centrifuge tube containing 5 mL of TB3 medium containing 10–50 μg/mL ampicillin, and incubate on a rotary shaker (160 rpm) at room temperature for 18–20 h (*see* **Note 11**).
6. Collect the regenerated mycelium by centrifugation at 2500 × *g* for 10 min, and suspend them in 500 μL of TB3 medium.
7. Add the regenerated mycelium in TB3 to 15 mL of melted bottom agar (~55 °C) containing 100 μg/mL hygromycin and 20–50 μg/mL ampicillin in a 50 mL sterile conical centrifuge tube, and quickly mix well into a homogenous solution (*see* **Note 12**).
8. Pour the mixture into a Petri dish, and incubate at room temperature (25 °C) for 10–15 h.
9. Add a layer of 15 mL of top agar containing 150 μg/mL hygromycin and 20–50 μg/mL ampicillin to reduce the possibility of background colony growth (*see* **Note 12**).
10. Hygromycin-resistant transformants usually emerge on the surface of the selective media in 4–7 days.
11. Pick transformants using a sterile inoculating needle or toothpicks to 24-well plates containing M-100 agar media with at least 75 μg/mL hygromycin.
12. Allow the transformants to grow for several days at room temperature.
13. Inoculate individual transformants to 3 mL PDB, and allow to grow for 5 days on a shaker at 180 rpm at room temperature.
14. Extract total DNA of the transformants for screening and confirming desired fusion by PCR. The desired transformants should be further purified using single spores for 1 ~ 2 rounds.
15. Sequence the locus at the site of the fusion to confirm there are no undesired mutations (insertions or deletions of nucleotides) that could cause a frameshift in the downstream coding sequence.

3.5 Evaluation of Protein Expression and Subcellular Localization by Microscopy

1. An inverted agar method is used for investigating the subcellular localization of the protein(s) of interest (*see* **Note 13**).
2. Inoculate the desired *F. oxysporum* transformant(s) with the correct fusion into 50 mL YG media, and culture for ~4 days at room temperature on a rotary shaker (150 rpm).
3. Use 500 μL of the hyphae-conidia mixture of each of the transformants, and inoculate to new tube of 50 mL YG media. Incubate these cultures at 18 °C on a shaker (200 rpm) for 16–18 h (*see* **Note 14**).
4. Extract 1 mL of the fungal culture, and quickly pellet the mycelia by centrifugation and remove the supernatant. Add 500 μL of 4% paraformaldehyde for cell fixation, and allow to stand for ~10 min at room temperature.
5. Wash the hyphae-conidia mixture twice using PBS (pH 7.4), and suspend in 300 μL of PBS.
6. Aseptically cut the solidified 2.5% water agar into 0.5 cm × 0.5 cm pieces.
7. Place a piece of water agar on a glass slide and drop 10 μL of the fixed hyphae-conidia mixture on the surface of the water agar, and gently place a cover glass on the water agar.
8. Invert the whole "sandwich" structure and image under confocal microscopy (*see* **Note 15**).

4 Notes

1. As an alternative to 1 g $CaCl_2$, 1.32 g of $CaCl_2 \cdot 2\ H_2O$ can be used.
2. Construction of the donor DNA usually requires multiple fragments (at least 3 fragments) to be assembled into a single plasmid. To obtain a higher assembly efficiency, it is recommended to use a 25 bp overlap at the ends of the fragments for assembly of multiple (4 or more) fragments. Using a standard restriction enzyme digestion followed by ligation is not recommended since this conventional method is usually much more time-consuming.
3. It is recommended to not use a PCR fragment that would have the cleavage site directly in the middle of the fragment, as when it is analyzed by gel electrophoresis, the two cleavage products could co-migrate together.
4. The molar ratio of the Cas9 protein and the sgRNA should be approximately 1:1. If not, the cleavage efficiency could be compromised. The maximum transformation efficiency was usually obtained using a combination of 10 μg Cas9 proteins and 2 μg sgRNA. Additionally, if Cas9 Nuclease Reaction

Buffer is not available, the NEB CutSmart® Buffer can be substituted.

5. It is recommended to use a medium that promotes conidia production, and it is recommended to test the *F. oxysporum* isolate of interest. We have compared the conidia production in PDB, V8, and M-100 liquid media and found M-100 was usually sufficient to produce a large number of conidia.
6. *F. oxysporum* isolates usually grow rapidly in suitable culture environments like PDB media. When protoplasting, conidia should not be cultured for over 14 h, as older germlings are more difficult to use to generate the ideal number of protoplasts during treatment with the protoplasting enzyme solution.
7. Enzymatic treatments should not be over 4 h. Long-time processing in the enzymatic treatment may disrupt protoplasts. In addition, the shaking rate on the rotary shaker should not be high, and this could result in unwanted lysis of the protoplasts.
8. The protoplast quality is one of the most important points for Cas9 RNP-mediated transformation. Ideal protoplasts should have spherical fungal cells and not contain a large amount of hyphae debris and other impurities.
9. The final concentration of fungal protoplasts during the transformation is important for the efficiency. Too low or high of a concentration of protoplasts could contribute to transformation failure.
10. Cas9 RNP-mediated fungal transformation usually has a higher efficiency for genome editing, so it might not be necessary to use an abundant amount of donor DNA. Additionally, more donor DNA could lead to an increase in the number of ectopic transformants. For example, conventional *F. oxysporum* transformation usually requires 5 or more μg of donor DNA, and the use of only 1–3 μg of donor DNA in the transformation may lead to no or very few transformants on the selective media.
11. The use of ampicillin is able to reduce most bacterial contamination in case it occurs during the transformations. This is not necessary for the procedure if the antibiotic is not available, as long as sterile technique is used and the procedure is conducted in a biosafety cabinet.
12. The temperature of the melted bottom or top agar media should not be high, or regenerated mycelium could be killed by the heat. The processing time in this step should be short to avoid solidification of the agar within the tubes.
13. Different reporter tags used in the fusion have various molecular activities and are suitable for different methods for

characterization. For example, fluorescent proteins like GFP, RFP, and mCherry can be easily detected and evaluated under fluorescent microscopy, and the markers are usually used for subcellular localization studies on native proteins. LacZ or GUS molecular markers show a color change reaction when the appropriate substrate is present and are usually used to investigate gene regulation and transcription.

14. The temperature and culture environment, including nutrients present, can influence fungal development. Unfavorable culture conditions may lead fungi to producing large vacuoles.
15. For live cell imaging, the fungi can be cultured in a four-chambered coverglass (Thermo Scientific™ Nunc™ Lab-Tek™ Chambered Coverglass) with a very thin layer of solidified culture media like PDA and M-100 agar.

Acknowledgments

Development of this method and publication was partially supported by the Alabama Agricultural Experiment Station, the Hatch program of the National Institute of Food and Agriculture, USDA, and the Special Project of Technological Innovation Guidance in Shaanxi Province (No. 2020QFY12-06).

References

1. Gordon TR (2017) *Fusarium oxysporum* and the Fusarium wilt syndrome. Annu Rev Phytopathol 55:23–39
2. Wang Q, Cobine PA, Coleman JJ (2018) Efficient genome editing in *Fusarium oxysporum* based on CRISPR/Cas9 ribonucleoprotein complexes. Fungal Genet Biol 117:21–29
3. Wang Q, Coleman JJ (2019) CRISPR/Cas9-mediated endogenous gene tagging in Fusarium oxysporum. Fungal Genet Biol 126:17–24
4. Wang Q, Coleman JJ (2019) Progress and challenges: development and implementation of CRISPR/Cas9 technology in filamentous fungi. Comput Struct Biotechnol J 17:761–769
5. McCluskey K, Wiest A, Plamann M (2010) The fungal genetics stock center: a repository for 50 years of fungal genetics research. J Biosci 35:119–126
6. Stevens RB (1974) Mycology guidebook. University of Washington Press, Seattle
7. Stemmer M, Thumberger T, del Sol KM, Wittbrodt J, Mateo JL (2015) CCTop: an intuitive, flexible and reliable CRISPR/Cas9 target prediction tool. PLoS One 10:e0124633

Chapter 9

In Vivo Monitoring of Cytosolic pH Using the Ratiometric pH Sensor pHluorin

Tânia R. Fernandes, Antonio Serrano, and Antonio Di Pietro

Abstract

Cytosolic pH (pH_{cyt}) is a key factor controlling cell fate. The genetically encoded pH-sensor pHluorin has proven highly valuable for studies on pH_{cyt} in many living organisms. pHluorin displays a bimodal excitation spectrum with peaks at 395 nm and 475 nm, which is dependent on pH. Here we describe two different protocols for determining pH_{cyt} in the soil-borne fungal pathogen *Fusarium oxysporum*, based either on population or single-cell analysis.

Key words Cytosolic pH, pHluorin, Ratiometric pH sensor, Spectrofluorometer, Confocal microscope

1 Introduction

Cytosolic pH (pH_{cyt}) is a critical factor of the intracellular environment determining life or death in all biological systems. Changes in pH_{cyt} affect the ionization state of a wide array of molecules including peptides, proteins, and lipids, which regulate fundamental processes in the cell [1]. Therefore, pH_{cyt} tends to be constant and tightly regulated in all organisms [2]. Nevertheless, rapid and transitory changes in pH_{cyt} can occur in response to different stimuli such as shifts in ambient pH or nutrient status [3–5]. It is increasingly appreciated that fluctuations in pH_{cyt} act as signals regulating key functions including cell growth and proliferation [6], life span [7], nutrient response [8], and pathogenicity [9].

An important step forward in understanding pH_{cyt} homeostasis has been the development of genetically encoded ratiometric pH-sensors. The pH-sensitive green fluorescent protein (GFP) derivative pHluorin has been used as a valuable tool to study pH_{cyt} in yeast [7, 8, 10–12], filamentous fungi [3, 13], and higher

Jeffrey Coleman (ed.), *Fusarium wilt: Methods and Protocols*,
Methods in Molecular Biology, vol. 2391, https://doi.org/10.1007/978-1-0716-1795-3_9,

eukaryotes [14]. Originally developed by Miesenbock and colleagues, pHluorin displays a bimodal excitation spectrum with peaks at 395 nm and 475 nm and an emission maximum at 510 nm. With intracellular acidity, pHluorin shows a dose-dependent decrease in excitation at 395 nm with a concomitant increase at 475 nm, whereas the opposite effect is observed at high pH_{cyt} [15]. These characteristics make pHluorin an ideal tool to precisely measure pH_{cyt} in any living organism.

Different approaches for in vivo monitoring of pH_{cyt} in pHluorin expressing strains have been developed. The most commonly used method is fluorescence microscopy, which allows the study of pH_{cyt} compartmentalization within the cell and offers spatial and temporal insights into population variability [3, 16]. Ratiometric flow cytometry analysis of pHluorin expressing cells combines high-throughput analysis at the single-cell level with monitoring of pH_{cyt} distribution within large cell populations [13, 17, 18]. Finally, analysis of pH_{cyt} in a microwell reader coupled to a spectrofluorometer has proven useful for high-throughput analysis, combining fast data acquisition with the possibility to simultaneously assay different experimental conditions [10].

In this chapter, we describe two different protocols for monitoring pH_{cyt} in the soil-borne fungal pathogen *Fusarium oxysporum*, based either on population or single-cell analysis. Previously, a pHluorin-expressing strain of the *F. oxysporum* f. sp. *lycopersici* isolate Fol4287 was obtained by co-transformation of protoplasts with a construct containing the *pHluorin* gene and the *Saccharomyces cerevisiae adh5* terminator [15] fused to the strong constitutive *Aspergillus nidulans gpdA* promoter [19], together with the *hygromycin* resistance (*Hyg*^{R}) cassette. Hygromycin-resistant transformants were screened by fluorescence microscopy, and transformants displaying the strongest fluorescence intensity were selected. In vivo calibration with buffers at different pH values confirmed the pH sensitivity and spectral characteristics of the expressed pHluorin, allowing the measurement of pH_{cyt} in this fungal pathogen [20].

2 Materials

Prepare all buffers and analytical grade reagents using ultrapure water (prepared by purifying deionized water, to attain a sensitivity of 18.2 MΩ-cm at 25 °C) and media with deionized water. Prepare and store all reagents at room temperature unless indicated otherwise. Diligently follow all waste disposal regulations when disposing waste materials.

2.1 Fungal Growth and Inoculation

1. Potato dextrose broth (PDB) medium: boil 200 g of peeled potatoes in 0.6 L of deionized water for 1 h. Stir and add 20 g of glucose (2%) (w/v). Add deionized water up to 1.0 L. Autoclave.
2. Autoclaved empty 125 mL Erlenmeyer flasks.

2.2 Measurement of Cytosolic pH in a Fungal Cell Population

1. Yeast extract Dextrose (YD) medium: 1% (w/v) glucose, 1% (w/v) yeast extract. Adjust pH to 8.0 with 1 M NaOH before autoclaving.
2. 1 M HEPES (4-(2-hydroxyethyl)-1-piperazineethanesulfonic acid) in ultrapure water, adjust pH to 7.4 with 10 M NaOH, and sterilize by filtration (0.22 μm pore size).
3. Autoclaved empty 250 mL Erlenmeyer flasks.
4. Sterile tubes and plastic centrifuge tubes. Autoclaved ultrapure water.
5. High speed tabletop centrifuge.
6. Thoma cell counting chamber, 0.100 mm (Paul Marienfeld GmbH & Co. KG).
7. pH calibration buffer: 50 mM HEPES, 50 mM MES, 50 mM KCl, 50 mM NaCl, 0.2 M ammonium acetate ($C_2H_7NO_2$), 10 mM NaN_3. Adjust the pH of different aliquots from 5.0 to 8.5, with 0.5 units steps. Sterilize by filtration (0.22 μm pore size), and add freshly prepared nigericin sodium salt (37 mg/ mL; Sigma Aldrich) to each aliquot of calibration buffer. Calibration buffer aliquots can be stored at 4 °C for a maximum of 4 weeks.
8. KSU buffer: 50 mM K_2HPO_4, 50 mM sodium succinate, 25 mM urea. Adjust pH to 6.0 with 1.2 M HCl.
9. Leica DMR Fluorescence Microscope.
10. 96-well microtiter plate.
11. Spectrofluorometer.

2.3 Single-Cell Analysis of Cytosolic pH

1. **Items 1–8** as described in the previous section.
2. Microscopy well slides (μ-Slide 8 well).
3. Laser confocal microscope Zeiss LSM880 on an inverted Axio Observer 7 (Carl Zeiss Microscopy GmbH), using objectives Plan Apo 63× oil 1.4 NA. Lasers used: diode (405 nm) and Argon (488 nm).

3 Methods

Carry out all procedures at room temperature under sterile conditions unless otherwise specified.

3.1 Fungal Growth and Inoculation

1. Inoculate around 10^6 microconidia of the *F. oxysporum* wild type and pHluorin-expressing strains in 25 mL PDB medium in a 125 mL Erlenmeyer flask (*see* **Note 1**). Incubate flasks for 3 days at 28 °C with orbital shaking at 170 rpm.
2. Collect microconidia by filtration through an autoclaved nylon filter (mesh size 10 μm), and centrifuge at 12,000 × *g* for 10 min.
3. Resuspend the microconidia in 1 mL of ultrapure water.
4. Determine the concentration of microconidia/mL by counting a 10^{-2} or 10^{-3} dilution in a Thoma cell counting chamber.
5. Inoculate 5 × 10^6 microconidia/mL in 50 mL YD pH 8.0 buffered with 1 mL of 1 M HEPES pH 7.4 (20 mM final concentration) (*see* **Note 2**). Incubate flasks for 15 h at 28 °C with orbital shaking at 170 rpm.

3.2 Measurement of Cytosolic pH in a Fungal Population

1. Collect germlings by filtration using an autoclaved nylon filter (mesh size 10 μm) or cheesecloth and vacuum filtration equipment. Discard flowthrough with ungerminated conidia.
2. Wash the germlings retained in the filter three times with 200 mL KSU buffer. In the last washing step, avoid letting the germlings dry completely, maintaining a small volume of the KSU wash buffer.
3. Resuspend germlings in KSU buffer at a concentration of 2.5 × 10^7/mL and vortex to mix.
4. For pH calibration, add 80 μL of the different calibration buffer aliquots ranging from pH 5.0 to 8.5 to the wells of a 96-well microtiter plate. Prepare triplicates for each strain (wild type and transformant expressing pHluorin; Fig. 1).
5. For measurement of pH_{cyt} in living cells, add 80 μL of KSU buffer to the wells of a 96-well microtiter plate. Prepare triplicates for each strain (Fig. 1).
6. Add 20 μL of the germling suspension in KSU (*see* **step 3**) to each well. Optical density at 600 nm should be 0.8–0.9.
7. Incubate the 96-well microtiter plate for 30 min at 28 °C with orbital shaking at 170 rpm (*see* **Note 3**).
8. Measure fluorescence at excitation wavelengths of 395 and 475 nm on a microplate spectrofluorometer every 5 min for 30 min (*see* **Note 4**).
9. Remove the microplate from the spectrofluorometer, and add the desired compound(s) (*see* **Note 5**) to the wells containing the germlings in KSU buffer.
10. Measure fluorescence at excitation wavelengths of 395 and 475 nm every 5 min for 1–2 h (*see* **Note 6**).
11. Export data as an Excel sheet.

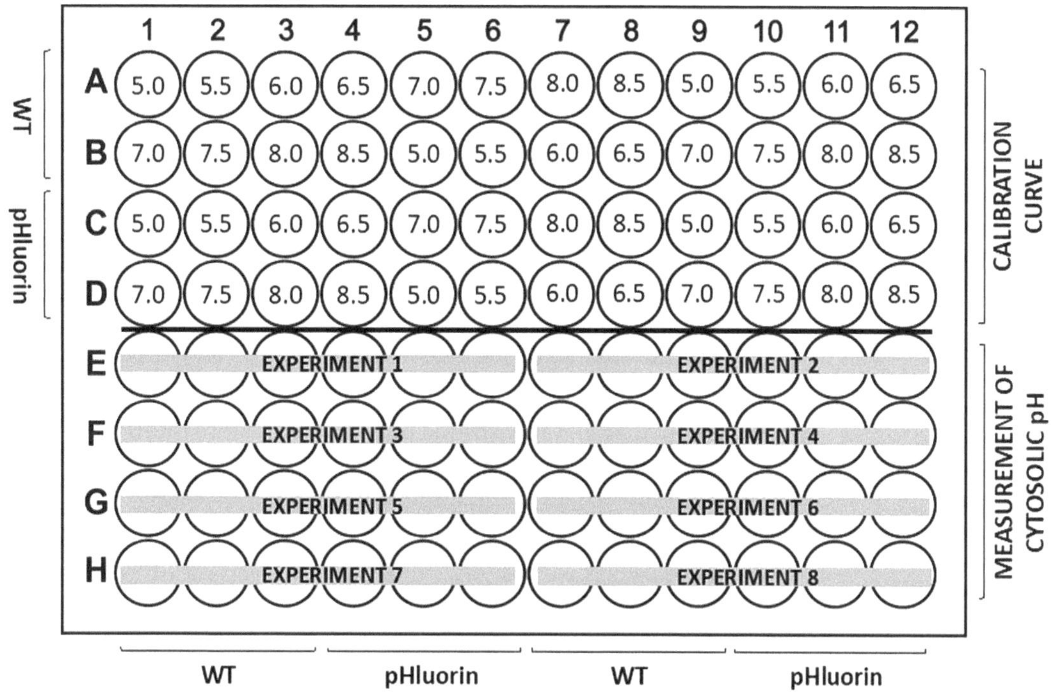

Fig. 1 Schematic diagram of the distribution of the wild-type (WT) strain and transformant expressing pHluorin in a 96-well microplate for the measurement of cytosolic pH (pH_{cyt}). For pH calibration, add 80 μL of the different calibration buffer aliquots ranging from pH 5.0 to 8.5 to each well (Rows A-D). For measurement of pH_{cyt} add 80 μL of KSU buffer to each well (Rows E-H). To start the assay, add 20 μL of the germling suspension in KSU to each well. For each strain, prepare triplicates for the calibration curve and each experimental sample as indicated

3.3 Single-Cell Analysis of Cytosolic pH

1. Perform **steps 1** and **2** as described in the previous section.
2. Resuspend germlings in KSU buffer at a concentration of 5×10^6 germlings/mL and vortex to mix.
3. Prepare a 10^{-3} dilution.
4. Add 200 μL to a microscopy microwell slide (around 1000 germlings/well).
5. Incubate the microwell slide for 30 min at 28 °C (*see* **Note 3**).
6. In the fluorescence microscope, set the excitation wavelength to 405 nm with the diode laser and to 488 nm with the Argon laser (preferentially use wavelengths in independent tracks). Set the emission wavelength to 510 nm (505–515 nm range) for both excitation wavelengths, and use the most sensitive detector (*see* **Note 7**).
7. Add desired test compounds to the microwell slide, either before or during image acquisition.

3.4 Measurement of Cytosolic pH

3.4.1 Calibration Curve

1. For each excitation wavelength and pH value of the calibration curve, subtract the background value (average of the triplicates of the wild-type strain) from the fluorescence values measured with the pHluorin expressing strain. For microscopy images, set image to 8 bits, and subtract the background using the lookup table HiLo (ImageJ/Fiji) and adjusting the contrast. Use the same background value for all images of an experiment.
2. Calculate the 395/475 nm ratio for each pH value of the calibration curve (Fig. 2a). For microscopy images, draw a line delimiting the shape of the cell, measure the intensity within the line for each wavelength, and calculate the 405/488 nm ratio for each pH value of the calibration curve.
3. Normalize all values to the mean 395/475 nm or 405/488 nm ratio obtained at pH 7.
4. Convert ratios of fluorescence intensity to pH_{cyt} values by fitting Eq. 1 below to the calibration curve (Fig. 2b). Determine pK_a as well as R_{min} and R_{max} for each set of experiments (*see* **Note 8**).

$$pH = pK_a - \log_{10} [(R_i - R_{min})/(R_{max} - R_i)] \qquad (1)$$

R_{max} and R_{min} are the limiting values for the ratio at the extreme acid (pH 5.0) and alkaline (pH 8.5) pH values, respectively, and pK_a is the apparent pK_a value of the pHluorin fluorophore.

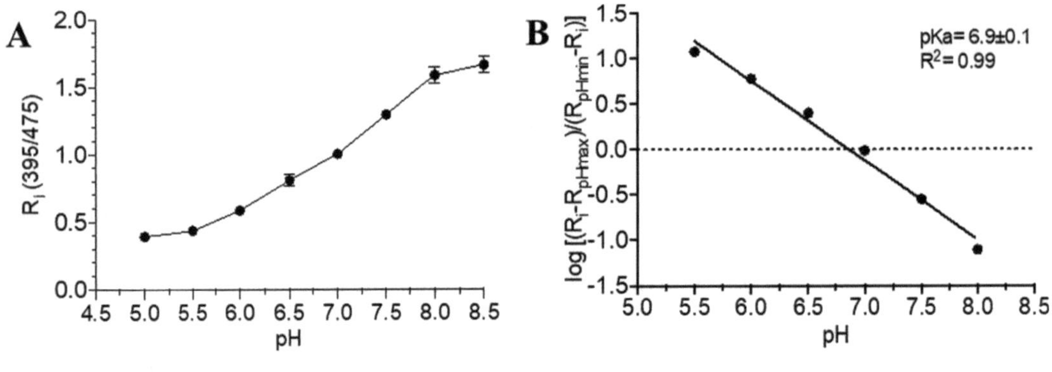

Fig. 2 Example of a calibration curve for the measurement of cytosolic pH (pH_{cyt}). (**a**) Plot of fluorescence excitation ratio (R_i) versus buffer pH. Germinated microconidia of *F. oxysporum* expressing pHluorin were washed and suspended in pH-adjusted calibration buffers ranging from 5.0 to 8.5, containing nigericin to equalize the extracellular pH with the pH_{cyt}. Excitation at 395 nm and 475 nm was measured, and the ratio between the two emission intensities collected at 510 nm was calculated. Data are shown as mean and standard deviation from at least three independent experiments, with three independent replicates for each condition. (**b**) Logarithmic plot of fluorescence excitation ratio versus buffer pH for calculation of the pKa constant. R_{max} and R_{min} are the limiting values for the ratio at the extreme acid (pH 5.0) and alkaline (pH 8.5) condition. R_i is the ratio for each pH determined in (**a**)

5. Draw a graph depicting pH_{cyt} versus the ratios for each pH value, and use the generated straight-line equation to calculate the precise value of pH_{cyt} in the different experimental samples (see below).

3.4.2 Experimental Samples

1. Calculate the 395/475 nm ratio by first subtracting the background value (average of the triplicates of the wild-type strain) in germlings resuspended in KSU buffer. For microscopy images, subtract the background for each experimental condition as described in the previous section, and calculate the 405/488 nm ratio.
2. Normalize all values to the mean 395/475 nm or 405/488 nm ratio obtained at pH 7.
3. Calculate the precise value of pH_{cyt} by applying the straight-line equation generated from the calibration curve (*see* **step 5** of the previous section).

4 Notes

1. Add hygromycin B at 20 μg/mL to the PDB medium to grow the wild-type expressing pHluorin strain (Hyg^R strain).
2. We recommend inoculating 25–50 mL of medium to obtain sufficient germlings for the experiment.
3. Incubate the microplate for 30 min to allow nigericin to equilibrate the intra- and extracellular pH for establishment of the calibration curve and to allow germlings to adapt to the KSU buffer for precise in vivo determination of pH_{cyt}.
4. Measurements were performed in an Infinite M200 PRO spectrofluorometer (TECAN Life Sciences). Software setup_Plate: Greiner 96 Flat Bottom Transparent Polystyrene Cat. No.655101/655161/655192 [GRE96ft.pdfx]/Target Temperature: 28 °C / Shaking (Linear) Duration: 3 s/Shaking (linear) Amplitude: 6 mm. Actions in the measurement script: 2 Labels – Mode: Fluorescence Top Reading/Multiple Reads per well (circle filled): 3 × 3/Multiple Reads per well (border): 750 μm/Excitation Wavelength: 395 nm or 475 nm/Emission Wavelength: 510 nm / Excitation Bandwidth: 9 nm/Emission bandwidth: 20 nm/Gain: 80 (Manual)/Number of flashes: 3/Integration Time: 40 μs/Lag Time: 0 μs/Settle Time: 0 ms/*Z*-Position (Manual): 20,000 μm.
5. Prepare the different compounds at a concentration of at least 10×.
6. Fluctuations of pH_{cyt} are generally rapid and transitory. However, measurements can be performed for hours or even days.

For longer measurements we recommend sealing the microplate with Parafilm and incubating in an incubator at the desired temperature.

7. Before acquiring the images, use three independent germlings resuspended in the calibration buffer at pH 6 to adjust the intensity of each laser, the gain of the detector, the exposure time, and the image resolution. The optimal settings will ensure short image acquisition times to avoid misreads or artefacts between the images acquired at the two excitation wavelengths, due to the natural cytoplasmic movement. Preferentially, use a single *Z* step adjusted to the center of the fungal cell. Once optimized, use the same settings for all images of a given experiment.
8. First determine $\log_{10}[(R_i - R_{min})/(R_{max} - R_i)]$ and draw a graph $\log_{10}[(R_i - R_{min})/(R_{max} - R_i)]$($x$ axis) versus the pH of the different calibration buffers (y axis). Determine the straight-line equation for this curve where pK_a will be the y-intercept value (Fig. 2b). Use pK_a and $\log_{10}[(R_i - R_{min})/(R_{max} - R_i)]$ values to calculate pH_{cyt}. The graph of pH_{cyt} versus ratio will be used as the calibration curve.

Acknowledgments

The authors acknowledge Prof. Zdena Palkova, Faculty of Science, BIOCEV, Charles University, Prague, for kindly sending the pHluorin plasmid and Maria Ortega Bellido for the valuable technical help. Research in the Di Pietro lab is supported by grants BIO2016-78923-R and PID2019-108045RB-I00 from the Spanish Ministerio de Ciencia e Innovación (MICINN). T.R.F. and A.S. were supported by the Marie Curie ITN FUNGIBRAIN (FP7-PEOPLE-ITN-607963).

References

1. Orij R, Brul S, Smits GJ (2011) Intracellular pH is a tightly controlled signal in yeast. Biochim Biophys Acta 1810:933–944
2. Kane PM (2016) Proton transport and pH control in fungi. Adv Exp Med Biol 892:33–68
3. Bagar T, Altenbach K, Read ND, Bencina M (2009) Live-cell imaging and measurement of intracellular pH in filamentous fungi using a genetically encoded ratiometric probe. Eukaryot Cell 8:703–712
4. Valkonen M, Mojzita D, Penttilä M, Benčina M (2013) Non-invasive high-throughput single-cell analysis of the intracellular pH of *Saccharomyces cerevisiae* by ratiometric flow cytometry. Appl Environ Microbiol 79:7179–7187
5. Ullah A, Lopes MI, Brul S, Smits GJ (2013) Intracellular pH homeostasis in *Candida glabrata* in infection-associated conditions. Microbiology (Reading) 159:803–813
6. Reshkin SJ, Greco MR, Cardone RA (2014) Role of pHi, and proton transporters in oncogene-driven neoplastic transformation. Philos Trans R Soc Lond Ser B Biol Sci 369:20130100
7. Hughes AL, Gottschling DE (2012) An early age increase in vacuolar pH limits

mitochondrial function and lifespan in yeast. Nature 492:261–265

8. Dechant R, Binda M, Lee SS, Pelet S, Winderickx J, Peter M (2010) Cytosolic pH is a second messenger for glucose and regulates the PKA pathway through V-ATPase. EMBO J 29:2515–2526
9. Fernandes TR, Segorbe D, Prusky D, Di Pietro A (2017) How alkalinization drives fungal pathogenicity. PLoS Pathog 13:e1006621
10. Isom DG, Sridharan V, Baker R, Clement ST, Smalley DM, Dohlman HG (2013) Protons as second messenger regulators of G protein signaling. Mol Cell 51:531–538
11. Peters LZ, Hazan R, Breker M, Schuldiner M, Ben-Aroya S (2013) Formation and dissociation of proteasome storage granules are regulated by cytosolic pH. J Cell Biol 201:663–671
12. Young BP, Shin JJ, Orij R, Chao JT, Li SC, Guan XL, Khong A, Jan E, Wenk MR, Prinz WA, Smits GJ, Loewen CJ (2010) Phosphatidic acid is a pH biosensor that links membrane biogenesis to metabolism. Science 329:1085–1088
13. Valkonen M, Penttilä M, Benčina M (2014) Intracellular pH responses in the industrially important fungus Trichoderma reesei. Fungal Genet Biol 70:86–93
14. Rathje M, Fang H, Bachman JL, Anggono V, Gether U, Huganir RL, Madsen KL (2013) AMPA receptor pHluorin-GluA2 reports NMDA receptor-induced intracellular acidification in hippocampal neurons. Proc Natl Acad Sci U S A 110:14426–14431
15. Miesenböck G, De Angelis DA, Rothman JE (1998) Visualizing secretion and synaptic transmission with pH-sensitive green fluorescent proteins. Nature 394:192–195
16. Orij R, Postmus J, Ter Beek A, Brul S, Smits GJ (2009) In vivo measurement of cytosolic and mitochondrial pH using a pH-sensitive GFP derivative in *Saccharomyces cerevisiae* reveals a relation between intracellular pH and growth. Microbiology (Reading) 155:268–278
17. Valli M, Sauer M, Branduardi P, Borth N, Porro D, Mattanovich D (2005) Intracellular pH distribution in Saccharomyces cerevisiae cell populations, analyzed by flow cytometry. Appl Environ Microbiol 71:1515–1521
18. Weigert C, Steffler F, Kurz T, Shellhammer TH, Methner FJ (2009) Application of a short intracellular pH method to flow cytometry for determining Saccharomyces cerevisiae vitality. Appl Environ Microbiol 75:5615–5620
19. Punt PJ, Oliver RP, Dingemanse MA, Pouwels PH, van den Hondel CA (1987) Transformation of Aspergillus based on the hygromycin B resistance marker from Escherichia coli. Gene 56:117–124
20. Serrano A (2014) *In vivo* determination of intracellular pH using pHluorin proteins in *Fusarium oxysporum*. Universidad de Córdoba, Spain

Chapter 10

Multi-Pronged Investigation of Volatile Compound-Mediated Interactions of *Fusarium oxysporum* with Plants, Fungi, and Bacteria

Ningxiao Li and Seogchan Kang

Abstract

Proteins and many biogenic compounds require water as a medium for movement. However, because volatile compounds (VCs) can travel through the air and porous soils due to their ability to vaporize at ambient temperature, they can mediate diverse intra- and inter-kingdom interactions and perform ecologically functions even in the absence of water. Here, we describe several tools and approaches for investigating how *Fusarium oxysporum* interacts with plants and other microbes through VCs and how VC-mediated interactions affect its ecology and pathology. We also present a method for capturing *F. oxysporum* VCs for analysis via gas chromatography linked to mass spectrometry.

Key words *Arabidopsis thaliana*, Chemical ecology, *Escherichia coli*, *Fusarium oxysporum*, GC-MS, I plate, Rhizosphere, Sandwiched plate assay, SPME, *Trichoderma*, Volatile compounds

1 Introduction

Fusarium oxysporum is a cosmopolitan soilborne fungal species complex that occupies diverse ecosystems and interacts with many organisms. Some members cause wilt disease in various plant species, including food crops, ornamentals, and timber species [1, 2], while certain endophytic strains seem to protect plants from pathogen attack [3]. Some strains infect immuno-compromised humans [4]. Diverse genes and regulatory pathways needed for pathogenicity have been studied [5]. However, relatively little is known about how secreted metabolites, especially those that are volatile, participate in *F. oxysporum*-plant and *F. oxysporum*-microbe interactions.

Volatile compounds (VCs), typically characterized by low molecular weight, hydrophobicity, and high vapor pressure at ambient temperature, are produced by plants, animals, and microbes. Because of their ability to travel through the air and porous soils even without water as a medium, VCs perform diverse

Jeffrey Coleman (ed.), *Fusarium wilt: Methods and Protocols*,
Methods in Molecular Biology, vol. 2391, https://doi.org/10.1007/978-1-0716-1795-3_10,

vital functions within and across multiple kingdoms [6–9]. Plant VCs induced by herbivore or pathogen attack, including green leaf volatile alcohols, esters, aldehydes, terpenes, and aromatic compounds, play crucial roles in defense, such as attracting parasitoids for herbivore control and priming or inducing plant defense responses [10–13]. Animals use VCs as signals for a variety of purposes, such as locating food, avoiding threats, and finding mates [14]. Similarly, increasing evidence suggests that VCs produced by bacteria and fungi contribute to competing for nutrients and space, regulating growth and development within and across species, and communicating with other organisms [8, 15–18]. Numerous microbial VCs have been found to alter plant growth, development, or physiology and help protect plants from biotic and abiotic stresses [19–21]. However, there have been limited attempts to investigate how *F. oxysporum* VCs contribute to its interaction with plants and other microbes [22–26], leaving the significance of VCs in the ecology and pathology of *F. oxysporum* vastly underappreciated.

We provide protocols that have been employed to investigate how *F. oxysporum* interacts with plants, fungi, and bacteria via VCs [22, 25, 27, 28]. *Arabidopsis thaliana*, *Trichoderma virens*, and *Escherichia coli* are used to illustrate how these protocols work. We also describe a protocol for capturing *F. oxysporum* VCs to be analyzed through gas chromatography linked to mass spectrometry (GC-MS). Comparative analysis of VCs produced by multiple strains via GC-MS, combined with results from assays on how their VCs affect other organisms, helps identify bioactive compounds. These protocols can be deployed to study diverse forms of VC-mediated interactions between other organisms with little or no modification.

2 Materials

2.1 Biological Materials

1. Seeds of *Arabidopsis thaliana* ecotype Col-0 (Lehle Seed Co.).
2. *Escherichia coli* BW25113.
3. *Trichoderma virens* G-41.
4. *Fusarium oxysporum* NRRL26379 and NRRL54003. Bacterial cells and fungal spores in 20% glycerol are stored in −80 °C freezer and are revitalized before experiments.

2.2 Culture Media and Solutions

1. Sucrose solution: Dissolve 500 g sucrose in 500 mL ddH_2O, adjust the volume to 1 L, and sterilize through filtration (0.2 μm filter).
2. Murashige and Skoog (MS) agar: 4.5 g MS basal medium and 7.5 g granulated agar in 1 L ddH_2O. Autoclave for 15 min at 121 °C on the liquid cycle. Add 5 mL sucrose solution (see above) after cooling to 50 °C.

3. 0.5× Potato Dextrose Agar (PDA): 19.5 g PDA in 1 L ddH_2O. Autoclave for 15 min at 121 °C on the liquid cycle.
4. 0.75× PDA: 29.25 g PDA in 1 L ddH_2O. Autoclave for 15 min at 121 °C on the liquid cycle.
5. Luria-Bertani (LB) broth: 25 g LB in 1 L ddH_2O. Autoclave for 15 min at 121 °C on the liquid cycle.
6. LB agar: 25 g LB and 10 g granulated agar in 1 L ddH_2O. Autoclave for 15 min at 121 °C on the liquid cycle.
7. 70% and 95% ethanol.
8. 6% sodium hypochlorite solution (CLOROX).
9. Dimethyl sulfoxide (DMSO).
10. Pro-Mix (Premier Tech Horticulture).
11. Hoagland solution [29]: 5 mL 1 M KNO_3, 2.5 mL 1 M KH_2PO_4, 2 mL 1 M $MgSO_4$, 2 mL 1 M $Ca(NO_3)_2$, 1 mL micronutrient mix (2.86 g H_3BO_3, 1.81 g $MnCl_2 \cdot 4H_2O$, 0.22 g $ZnSO_4 \cdot 7H_2O$, 0.08 g $CuSO_4 \cdot 5H_2O$, and 0.02 g $H_2MoO_4 \cdot H_2O$ in 1 L ddH_2O), and 1 mL Fe-EDTA solution (10.4 g EDTA-Na_2, 7.8 g $FeSO_4 \cdot 7H_2O$, and 56.1 g KOH in 1 L ddH_2O) in 1 L ddH_2O (*see* **Note 1**).

2.3 Equipment and Supplies

1. Surgical blades (#10 and #11).
2. Cork borers (#2 and #6).
3. Cellophane membrane (Paper Mart, Orange, CA) disks (85 mm in diameter; *see* **Note 2**).
4. 100 × 15 mm I plate.
5. 100 × 15 mm Petri plate.
6. 100 × 100 mm square plate.
7. 24-well tissue culture plate.
8. 100 × 40 mm culture plate.
9. Solid-phase microextraction (SPME) fiber assembly with 50/30 μm DVB/CAR/PDMS fiber coating (Supelco, Bellefonte, PA) attached to SPME holder (Supelco).
10. Flexible-arm electrode holder.
11. 25 °C incubator.
12. Growth chamber.
13. Spectrophotometer.

2.4 Homemade System for Assaying Bioactivities of Fungal VCs (Fig. 1; See Note 3)

1. Plastic deli container with lid (8 oz., top, 115 mm in diameter; bottom, 95 mm in diameter, 24 mm in height).
2. Silicone tubing (10 mm OD, 6 mm ID).
3. Flat washers (M10 Bolt, Nylon, 22 mm OD).

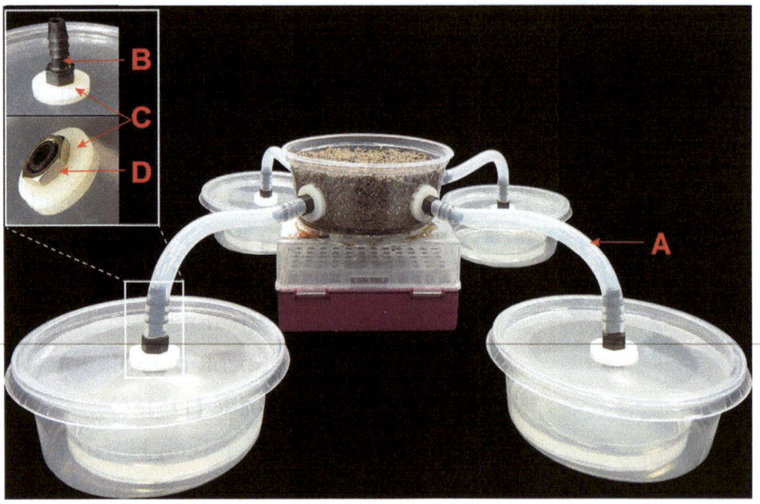

Fig. 1 A homemade system for assaying how *F. oxysporum* VCs affect plants grown in soils, the assembly of soil/rhizosphere microbiomes, or individual soil microbes. The container positioned at the center is connected through silicone tubing (**a**) to four containers, with each housing a single fungal culture plate. The top and bottom of the joint used to connect these containers are shown in the insert. Each joint consists of a male adaptor (**b**), two flat washers (**c**), and one hex panel nut (**d**)

4. Hex panel nuts (1/8″ in diameter, 1/8″ in height).
5. Male adapters (1/4″ barb size, Nylon).

3 Methods

3.1 I Plate-Based Assay for Investigating How F. oxysporum VCs Affect the Growth and Development of A. thaliana

The assay illustrated in Fig. 2 is simple to set up and employs the I plate. The center divider separates the I plate into two compartments so that the VC-mediated interaction between microbes and plants can be studied without their physical contact and water-mediated molecular exchange. Although the assay has some limitations (*see* Subheading 3.2), it enables quick screening of a large number of fungal or bacterial strains. By adjusting the concentration or type of individual nutrients in the fungal culture medium, one can also investigate if specific nutrients or their availability influence the production of VCs that affect plants. We employed this assay to investigate how *F. oxysporum* VCs affect plant growth and development [22] and how *A. thaliana* exposed to *F. oxysporum* VCs responds to stresses [28].

1. Inoculate ~20 μL glycerol stock of *F. oxysporum* spores at the center of a 0.5× PDA plate, and incubate at 25 °C for 7 days.
2. Surface sterilize *A. thaliana* seeds by washing with 95% ethanol for 1 min and 6% sodium hypochlorite solution for 15 min.

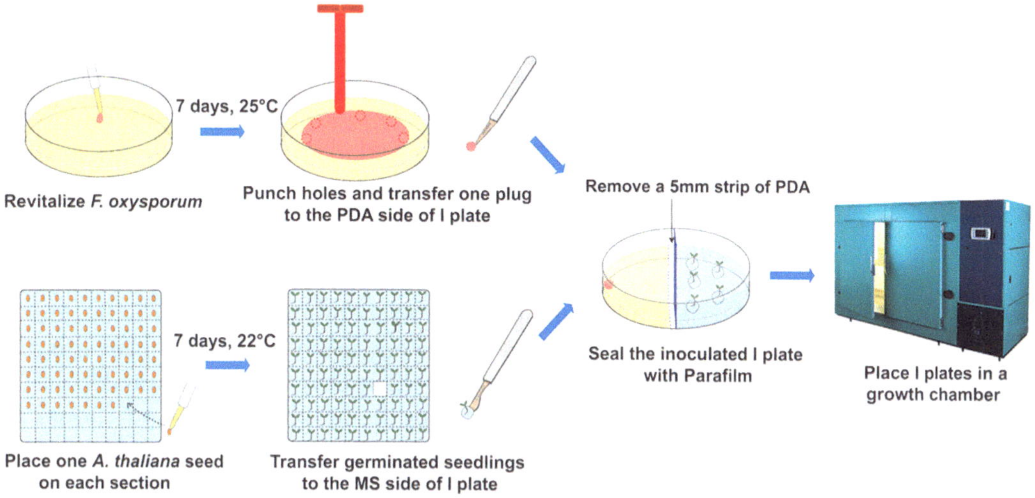

Fig. 2 The workflow of the treatment of *A. thaliana* seedlings with *F. oxysporum* VCs using an I plate

Wash twice with sterile ddH_2O before incubating the seeds in sterile ddH_2O at 4 °C in the dark for 3 days.

3. Slice MS agar in square Petri plate into 10 × 10 mm pieces using a sterilized blade # 10.
4. Sow one seed on each section of MS agar using a 10 μL micropipette. Seal the plate with two layers of Parafilm, and place them in a plant growth chamber for 7 days. Set the chamber at 22 °C, 12 h light (4500 lux, 60 μmol photons m^{-2} s^{-1}), and 60% relative humidity.
5. Add 8 mL of MS agar to one compartment of the I plate and 8 mL of 0.5× PDA to the other. Leave the prepared plates in a laminar flow hood overnight before use to prevent excessive condensation.
6. Remove a 5-mm-wide strip of PDA along the center divider to prevent *F. oxysporum* hyphae from growing over to the MS side.
7. Transfer five seven-day-old *A. thaliana* seedlings (similar in size and growth stage) along with the attached agar piece to the MS side using a sterilized blade #10.
8. Punch holes along the actively growing margin of a *F. oxysporum* culture using a flame sterilized cork borer # 2, and transfer one plug (mycelial side down) to the far end of the PDA side of the I plate using a sterilized blade #11.
9. Seal the inoculated I plates with two layers of Parafilm, and place them in a plant growth chamber under the conditions described above for up to 14 days.
10. Seedlings of *A. thaliana* treated with *F. oxysporum* VCs are analyzed as described in Subheading 3.3 to determine how the VCs affect their physiology, growth, and development.

3.2 Determination of How *F. oxysporum* VCs Affect *A. thaliana* Grown in Soils

The I plate-based assay described above has a few challenges/limitations. One big challenge is how to separate the CO_2 effect on plants from that caused by other VCs [30]. In this small closed system, the rapid increase in CO_2 concentration resulting from fungal respiration causes enhanced plant growth [22, 30]. Capturing CO_2 using KOH or $Ba(OH)_2$ solution has been used [22, 30], but it is not an ideal solution. Plant VCs may affect the production of fungal VCs and consequently influence plant responses. This assay is not designed to test this possibility but simply reveals the outcome of signaling events likely originating from both sides of the interaction. One technical limitation is that the I plate can only contain small plants for up to ~2 weeks due to the limited space (*see* **Note 4**). Besides, because both the plant roots and shoots are exposed to microbial VCs, the assay is not suitable for determining which part(s) of the plants perceive the volatile signals. Since *F. oxysporum* is a soilborne fungal pathogen, an experimental system limiting VC exposure to the roots is highly desirable. A system that permits studies on how fungal VCs affect plants grown in soils is also needed. Here, we present a soil-based system (Fig. 3). This system supports inquiries about if and how *F. oxysporum* VCs contribute to its belowground interaction with *A. thaliana* or other plants (*see* **Note 5**) without concerns about changes caused by physical contact or other signals (e.g., secreted proteins or metabolites that move using water as a medium). Another important potential application of this system is described in **Note 5**.

1. Revitalize *F. oxysporum* and surface sterilize *A. thaliana* seeds, as described in Subheading 3.1.
2. Sow three seeds in each well of a 24-well plate filled with sterilized Pro-Mix (autoclaved twice before use).
3. Place the 24-well plate inside a clear container to prevent dehydration and incubate in a plant growth chamber under

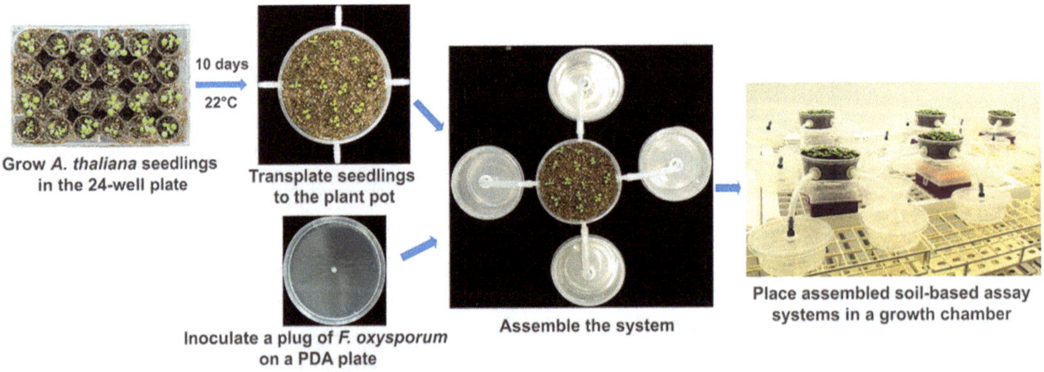

Fig. 3 The workflow of the soil-based assay system used to characterize how *F. oxysporum* VCs affect *A. thaliana*

the conditions described in Subheading 3.1. Water every 2 days with Hoagland solution and water, alternatively, for 10 days.

4. Thin *A. thaliana* seedlings to one per well using forceps and transplant 13 seedlings (similar in size and growth stage) along with the soil attached to the roots to each pot filled with sterilized Pro-Mix.
5. Connect the pot to four containers, with each container holding a plate of *F. oxysporum* cultured on PDA. The control consists of the plant pot connected with four containers that hold an un-inoculated PDA plate.
6. Place the assembled treatment systems in a plant growth chamber under the conditions described in Subheading 3.1, and water twice a week with Hoagland solution and water, alternatively, for 14 days.
7. Seedlings of *A. thaliana* treated with *F. oxysporum* VCs are analyzed as described in Subheading 3.3 to determine how the VCs affect their physiology, growth, and development (*see* **Note 5**).

3.3 Assessment of Various Changes in A. thaliana After Exposure to F. oxysporum VCs

Here, we briefly describe how to prepare plant samples for analyzing phenotypic or physiological changes caused by VC exposure using the procedures of measuring shoot and root weight and leaf chlorophyll content as examples. Collected plant samples can also be analyzed at the molecular level (e.g., transcriptome, proteome, metabolome) to investigate the mechanism underlying plant responses to microbial VCs [31, 32].

1. After co-cultivation (Fig. 4a), use forceps to remove *A. thaliana* plants from MS medium without damaging their roots.
2. Remove excess moisture on the collected plants using a paper towel.
3. After detaching the roots from the shoot, immediately weigh both samples using an analytical balance (*see* **Note 6**).
4. Measure leaf chlorophyll content [33]. After weighing all the leaves from each plant, place them in a sterile 15 mL Falcon tube containing 7 mL DMSO, and incubate them at 65 °C for 8 h. After transferring the chlorophyll extract to a graduated tube, add DMSO to bring the extract up to a total volume of 10 mL. Measure OD values at 645 and 663 nm in a spectrophotometer against a DMSO blank. The total chlorophyll content is calculated using the following formula:

$$\text{Total chlorophyll}\left(\text{mg g}^{-1}\right)=\left(\frac{20.2\times \text{OD}_{645}+8.02\times \text{OD}_{663}}{1000\times W}\right)\times V$$

Fig. 4 Growth responses of *A. thaliana* to VCs produced by *F. oxysporum* in two different assay systems. Plant growth in response to VCs produced by *F. oxysporum* strain NRRL 26379 was compared with that of the control (no *F. oxysporum* VCs). Representative data after 14 days of co-cultivation in (**a**) I plate (Fig. 2) and (**b**) soil-based assay system (Fig. 3) are shown

where *V* is the volume of DMSO and *W* is the fresh weight of leaves.

5. Plants treated with *F. oxysporum* VCs using the soil-based assay system (Fig. 4b) can be analyzed as described in **steps 1–4** with slight modifications (*see* **Note** 7).

3.4 Sandwiched Plate Assay Used to Study VC-Mediated Interactions Between F. oxysporum and T. virens

This assay was employed to show that VCs produced by *F. oxysporum* and four *Trichoderma* biocontrol agents affect each other in multiple ways [25]. We describe how to determine if VCs produced by *F. oxysporum* and *Trichoderma* affect the growth of each other (Subheadings 3.4.1 and 3.4.2) and whether *Trichoderma* and *F. oxysporum* can recognize the presence of each other by sensing the VCs from the other side (Subheadings 3.4.3 and 3.4.4).

3.4.1 Assessment of the Inhibitory Activity of T. virens VCs Against F. oxysporum

1. Revitalize *F. oxysporum* NRRL 26379 and *T. virens* G-41 by culturing them on 0.75× PDA for 7 and 3 days, respectively.
2. Inoculate one plug of *F. oxysporum* NRRL 26379 and *T. virens* G-41 cultures at the center of the 0.75× PDA plates.

Fig. 5 Sandwiched plate assay used to measure the antifungal activity of VCs produced by *T. virens*. The growth of *F. oxysporum* NRRL 26379 (top) after 5 days of co-cultivation with *T. virens* (bottom) was compared with that in control (no *T. virens* = un-inoculated PDA plate). VCs produced by *T. virens* significantly suppressed the growth of *F. oxysporum* compared to the control treatment

3. Place an *F. oxysporum* plate on top of a *T. virens* plate, and seal with three layers of Parafilm. The control consists of *F. oxysporum* plates sandwiched with un-inoculated PDA plates.
4. Incubate the sandwiched plates at 25 °C for 5 days (Fig. 5; *see* **Note 8**).
5. Measure the colony diameter of *F. oxysporum* to determine the degree of growth inhibition caused by *T. virens* VCs using the following formula:

$$\text{Degree of growth inhibition } (\%) = \left(\frac{D_{\text{control}} - D_{\text{treatment}}}{D_{\text{control}}}\right) \times 100$$

where D_{control} and $D_{\text{treatment}}$ indicate the colony diameters of *F. oxysporum* in the control and VC treatments, respectively.

3.4.2 Assessment of the Inhibitory Activity of F. oxysporum VCs Against T. virens

1. Revitalize *F. oxysporum* NRRL 26379 and *T. virens* G-41, as described in Subheading 3.4.1. The following steps are identical to those described in Subheading 3.4.1 with some modifications (e.g., the age of fungal culture used, co-cultivation time).
2. Transfer one plug of *F. oxysporum* NRRL 26379 to a 0.75× PDA plate, and incubate at 25 °C for 5 days to ensure enough fungal biomass.
3. Transfer one plug of *T. virens* to a 0.75× PDA plate, and immediately sandwich the inoculated plate with an *F. oxysporum* culture plate (on top). Seal with three layers of Parafilm. The control consists of *T. virens* plates sandwiched with un-inoculated PDA plates.
4. Incubate the sandwiched plates at 25 °C for 36 h (*see* **Note 8**).

5. Measure colony diameters of *T. virens* cultures to determine the degree of growth inhibition (%) caused by *F. oxysporum* VCs using the formula described in Subheading 3.4.1.

3.4.3 Assessment of VC-Mediated Recognition of F. oxysporum by T. virens (Fig. 6)

1. Revitalize *F. oxysporum* NRRL 26379 and NRRL 54003 and *T. virens* G-41, as described in Subheading 3.4.1.
2. Transfer one plug of NRRL 26379 and NRRL 54003 onto 0.75× PDA plates, and incubate at 25 °C for 5 days.
3. One day before setting up the sandwiched plate assay, place a cellophane membrane disk on each 0.75× PDA plate (*see* **Note 9**).
4. After transferring one plug of the *T. virens* culture onto the cellophane membrane overlaid on 0.75× PDA plate, immediately sandwich the inoculated plate with an *F. oxysporum* NRRL 26379 culture plate (the latter being on top). Seal the plates with three layers of Parafilm. The control consists of *T. virens* plates sandwiched with un-inoculated PDA plates.
5. After 40 h of incubation at 25 °C (*see* **Note 10**), disassemble the sandwiched plates.
6. Mark the spot where the plug of *T. virens* culture is located on the bottom of the plate using a permanent marker.
7. Remove the cellophane membrane along with the *T. virens* culture using two sterilized forceps (*see* **Note 11**).
8. Inoculate a culture plug of an indicator strain (a strain used to measure the antifungal activity of molecules secreted by *T. virens* in response to different treatments, e.g., *F. oxysporum* NRRL 54003) at the center of each *T. virens*-treated plate (*see* **step 6**).
9. Seal the plates with Parafilm and incubate at 25 °C for 5 days.

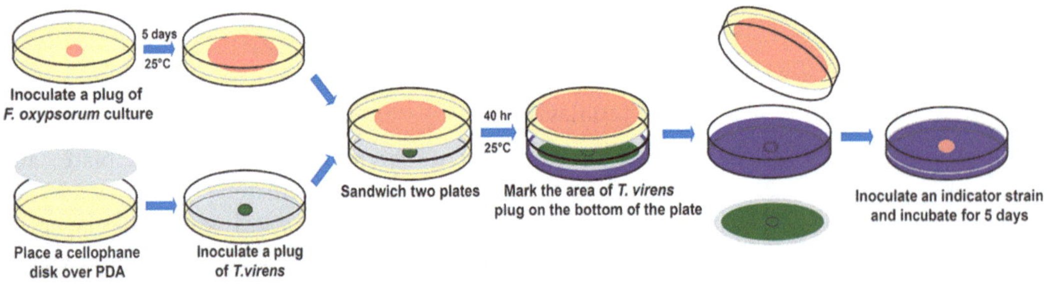

Fig. 6 Workflow of the sandwiched plate-based assay used to determine if *T. virens* senses VCs from *F. oxysporum*. Colony diameter of the indicator *F. oxysporum* isolate compared to that of control treatments (*T. virens* plate not exposed to *F. oxysporum* VCs) indicates if *F. oxysporum* VCs induce the quantity or activity of antifungal molecules secreted by *T. virens*

10. Measure the colony diameter of *F. oxysporum* NRRL 54003 to determine whether *F. oxysporum* VCs induced the quantity or activity of antifungal molecules secreted by *T. virens*.

3.4.4 Assessment of VC-Mediated Recognition of T. virens by F. oxysporum

1. Revitalize *F. oxysporum* NRRL 54003 and *T. virens* G-41, as described in Subheading 3.4.1. The following steps are identical to those described in Subheading 3.4.3 with some modifications (e.g., the age of fungal culture used, co-cultivation time).
2. Transfer one plug of the *T. virens* culture onto each 0.75× PDA plate, and incubate the inoculated plates at 25 °C for 1 day.
3. Inoculate one plug of the *F. oxysporum* culture on the cellophane membrane overlaid on each 0.75× PDA plate (*see* **Note 9**), and incubate at 25 °C for 3 days (*see* **Note 12**).
4. Sandwich each *F. oxysporum* plate with a *T. virens* plate (on top), and seal with three layers of Parafilm. The control consists of an *F. oxysporum* plate sandwiched with an un-inoculated PDA plate.
5. After 2 days of incubation at 25 °C, disassemble the sandwiched plates.
6. Mark the spot where the plug of the *F. oxysporum* culture was placed on the bottom of the plate using a permanent marker.
7. Remove the cellophane membrane along with the *F. oxysporum* culture using two sterilized forceps.
8. Inoculate a plug of *T. virens* culture at the same spot (*see* **step 6**) at the center of each *T. virens*-treated plate.
9. Seal the plate with Parafilm and incubate at 25 °C for 2 days.
10. Measure the colony diameter of *T. virens* to determine whether *T. virens* VCs induced the quantity or activity of antifungal molecules secreted by *F. oxysporum*.

3.5 Sandwiched Plate Assay to Study the VC-Mediated Interaction Between F. oxysporum and E. coli

We investigated how VCs produced by *Trichoderma* and tomato rhizosphere bacteria affect each other using the protocols described below [27]. This protocol was also applied to study VC-mediated interactions between *F. oxysporum* and rhizosphere bacteria. Subheadings 3.5.1 and 3.5.2 present the protocols used for determining the effect of VCs produced by *F. oxysporum* and *E. coli* on the growth of one another (*see* **Note 13**). Subheading 3.5.3 outlines the approach used to assess if and how *F. oxysporum* responds to VCs produced by *E. coli*.

1. Revitalize *E. coli* BW25113 on LB agar and *F. oxysporum* NRRL 26379 on a 0.75× PDA plate for 2 and 7 days, respectively.

Fig. 7 Sandwiched plate assay used for assessing if bacterial VCs affect the growth of *F. oxysporum*. A plate of two-day-old *E. coli* culture (top) sandwiched with a plate containing a plug of *F. oxysporum* culture (bottom) is shown

3.5.1 Assessment of the Inhibitory Activity of E. coli VCs Against F. oxysporum

2. Inoculate a single colony of *E. coli* to 2 mL of LB broth, and culture by shaking (200 rpm) at 25 °C until $OD_{600} = 1$ (equals 8×10^8 cells/mL).
3. Dilute the culture to a final concentration of 1×10^4 cells/mL, and plate a 100 μL cell suspension (~1×10^3 cells) to LB agar. Dry inoculated plates (without lids) in a laminar flow hood for 10 min (*see* **Note 14**).
4. After inoculating one plug of the *F. oxysporum* culture on each 0.75× PDA plate, immediately sandwich the plate with an *E. coli* plate (on top), and seal with three layers of Parafilm (Fig. 7). The control consists of a *F. oxysporum* plate sandwiched with an un-inoculated LB agar plate.
5. Incubate the sandwiched plates at 25 °C for 5 days (*see* **Note 8**).
6. Measure the colony diameter of *F. oxysporum* to determine the degree of growth inhibition (%) caused by *E. coli* VCs using the formula described in Subheading 3.4.1.

3.5.2 Assessment of the Inhibitory Activity of F. oxysporum VCs Against E. coli

1. Revitalize *E. coli* BW25113 and *F. oxysporum* NRRL 26379, as described in Subheading 3.5.1. The following steps are identical to those described in Subheading 3.5.1 with some minor modifications (e.g., the age of fungal cultures used, bacterial cell density, co-cultivation time).
2. Transfer one plug of *F. oxysporum* NRRL 26379 to each 0.75× PDA plate, and incubate at 25 °C for 5 days.
3. Cultivate *E. coli* in LB broth and then plate ~500 cells on LB agar. Dry inoculated plates (without lids) in a laminar flow hood for 10 min.

4. Sandwich a *F. oxysporum* plate with an *E. coli* plate (on top), and seal them with three layers of Parafilm. The control consists of an *E. coli* plate sandwiched with an un-inoculated PDA plate.
5. Incubate the sandwiched plates at 25 °C for 2 days (*see* **Note 8**).
6. Count the CFU to determine the degree of growth inhibition caused by *F. oxysporum* VCs using the following formula:

$$\text{Degree of growth inhibition (\%)} = \left(\frac{\text{CFU}_{\text{control}} - \text{CFU}_{\text{treatment}}}{\text{CFU}_{\text{control}}}\right) \times 100$$

where $\text{CFU}_{\text{control}}$ and $\text{CFU}_{\text{treatment}}$ indicate CFU in the control and VC treatments, respectively (*see* **Note 15**).

3.5.3 Assessment to Determine if Bacterial VCs Alter the Biosynthesis of Antibacterial Molecules Secreted by F. oxysporum

1. Revitalize *E. coli* BW25113 and *F. oxysporum* NRRL 54003, as described in Subheading 3.5.1.
2. Cultivate *E. coli* in LB broth, and then plate ~1×10^3 cells on LB agar, as described in Subheading 3.5.1.
3. Incubate the inoculated plates at 25 °C for 2 days to ensure enough bacterial biomass.
4. Inoculate one plug of the *F. oxysporum* NRRL 54003 culture on a cellophane membrane overlaid on each 0.75× PDA plate (*see* **Note 9**), and incubate them at 25 °C for 3 days (*see* **Note 12**).
5. Sandwich a *F. oxysporum* plate with an *E. coli* plate (on top), and seal them with three layers of Parafilm. The control consists of an *F. oxysporum* plate sandwiched with an un-inoculated LB plate.
6. After 2 days of incubation at 25 °C, disassemble the sandwiched plates.
7. Proceed with one of the following options to determine whether *E. coli* VCs induce the quantity or activity of antibacterial molecules secreted by *F. oxysporum*.

   Option 1:

   (a) Mark the area colonized by the *F. oxysporum* culture with a permanent marker, and then remove the cellophane membrane along with the *F. oxysporum* culture.

   (b) Plate ~500 *E. coli* cells on the *F. oxysporum* plate, and dry the inoculated plates (without lids) in a laminar flow hood for 10 min.

   (c) Incubate the plates at 25 °C for 2 days.

   (d) Count the number of CFUs within the marked area (*see* **Note 15**).

Option 2 (*see* **Note 16**):

(a) Add 5 mL of LB broth to the *F. oxysporum* plate after removing the cellophane membrane, and shake at 70 rpm for 3 h to allow secreted molecules to diffuse into LB.

(b) Transfer 1 mL of the LB broth containing the secreted molecules to a sterile 10 mL culture tube, and inoculate 10 μL of an *E. coli* culture (OD_{600} = ~1).

(c) Measure OD_{600} after overnight cultivation (200 rpm, 25 °C).

3.6 Extraction of VCs Produced by F. oxysporum for GC-MS Analysis

The following protocol for Solid Phase Microextraction (SPME) was employed to collect VCs produced by *F. oxysporum* for GC-MS analysis [25, 34]. Although capturing VCs via SPME is easy to perform and has been widely used, this method has some limitations (*see* **Note 17**). We already published a step-by-step protocol for separating and identifying individual VCs extracted using SPME via GC-MS [34]. More information about how to capture and analyze VCs can be found in our review on metabolomics [35].

1. Punch a hole on the lid (5 mm away from the edge) of a 100 × 40 mm culture plate using a flamed cork borer #6 (Fig. 8a). Disinfect the lid with 70% ethanol. Seal the hole with scotch tape.
2. Pour 250 mL of 0.75× PDA to each culture plate, and leave prepared plates in a laminar flow hood for 1 day to remove condensed water.
3. Remove a PDA block beneath the hole to make space for VC extraction (Fig. 8a).
4. Inoculate one plug of *F. oxysporum* culture at the center of each PDA plate, and seal it with three layers of Parafilm. An

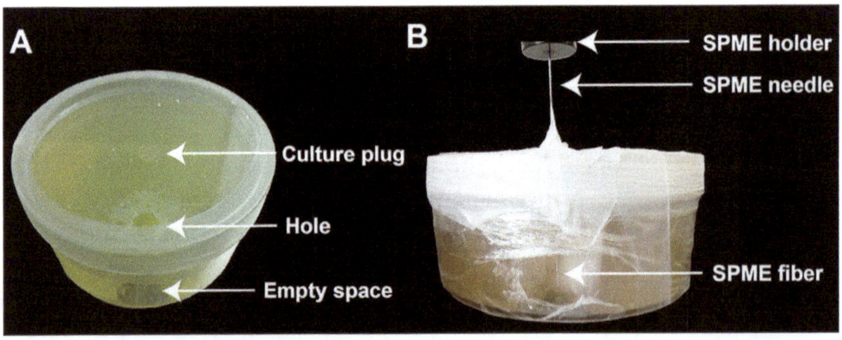

Fig. 8 Experimental setup for capturing *F. oxysporum* VCs using SPME. (**a**) The culture plate is modified by creating a hole on the lid (taped), and a block of PDA medium beneath the hole was removed to make space for inserting the SPME needle. A plug of an *F. oxysporum* culture is inoculated before sealing the plate. (**b**) The SPME needle is inserted into the headspace through the hole to extract *F. oxysporum* VCs. The SPME holder is attached to an electrode holder to secure the position of the needle

un-inoculated PDA plate serves as a control to help identify background VCs derived from the medium and the plate.

5. Incubate the plates at 25 °C for 14 days (*see* **Note 18**).
6. Condition the SPME fiber by placing the SPME needle into the GC injection port set at 230 °C for 1 h.
7. Insert the conditioned SPME needle through the hole into the headspace of the sampling plate. Use a flexible-arm electrode holder to lock the SPME needle position so that the insertion depth is consistent for all extractions.
8. Wrap the culture plate and the SPME needle with Parafilm to prevent VCs from escaping (Fig. 8b).
9. Extract VCs from each fungal sample for 1 h.

4 Notes

1. Prepare individual stock solutions separately, and sterilize them through filtration to prevent precipitation. For 1 L of Hoagland solution, add those solutions sequentially to 986.5 mL of autoclaved ddH_2O after cooling down to room temperature.
2. The cellophane membrane disk placed on top of the culture medium prevents fungal hyphae from reaching the medium but allows the passage of secreted molecules into the medium [25]. Stack cellophane membrane disks between filter papers, and soak in ddH_2O before autoclaving.
3. All parts, except plastic containers, are autoclaved at 121 °C for 15 min before assembly to minimize contamination. This step is critical for the experiments designed to study how fungal VCs affect soil/rhizosphere microbiomes (*see* **Note 5**). The plastic containers are sterilized using 70% ethanol. A loosely packed cotton plug is inserted in the silicone tubing connected to each joint of the plant pot/container to prevent soil contaminants from entering the fungal containers.
4. We have successfully assessed the effect of *F. oxysporum* VCs on *Nicotiana benthamiana* [22] and *Physcomitrella patens*, a moss species widely used as a model for plant evolution and development, using this assay. However, an assay with tomato seedlings was not feasible due to their large size and fast growth.
5. The system in Fig. 3 can be used to study how *F. oxysporum* VCs affect plant species larger than *A. thaliana* by replacing the pot with a larger pot (*see* Fig. 9). With different plants, the growth and treatment conditions may need to be modified. This system can also be used to investigate if *F. oxysporum* VCs affect rhizosphere/soil microbiomes or specific soil organisms (e.g., different pathogens, biocontrol agents). DNA extracted

Fig. 9 A system designed for studying how fungal VCs affect the assembly of tomato rhizosphere microbiomes. A pot (3.5″ × 3.5″ × 3″) with one 14-day-old tomato seedling is connected to four containers that house a single fungal culture plate. A control treatment (plant pot connected to un-inoculated PDA plates) is shown

from soil samples collected after individual treatments can be analyzed by metagenome sequencing or organism-specific quantitative PCR.

6. Keep the I plates and the plant pots covered while weighing harvested plants to minimize weight loss due to increased transpiration caused by airflow and low humidity in the laboratory. This step is critical for small plants like *A. thaliana* but may not be crucial for large plants.
7. It is challenging to harvest *A. thaliana* roots without physical damage and remove all attached soil particles. Given this challenge, we recommend that only aerial parts are analyzed. However, for larger plants like tomato (*see* Fig. 9), the roots can be easily separated from the bulk soil for analyses of morphological traits after exposure to fungal VCs.
8. Co-cultivation time depends on the growth rate of each fungal/bacterial strain to be tested. For example, 5 days is required for *F. oxysporum* to show any difference in colony size between treatments. As for *T. virens*, 36 h is chosen because its colony almost reaches the edge of the culture plate. It takes between 1 and 2 days for *E. coli* to form decent size colonies. Check the

growth rate of tested strains before setting up large-scale experiments.

9. Ensure that no air bubbles are trapped under the membrane because they result in uneven distribution of the secreted molecules that pass through the membrane. If needed, use a sterile spatula to push air bubbles toward the edge of the membrane. Leave the plates in a laminar flow hood overnight to remove excess moisture on the membrane.
10. Co-cultivation time depends on the growth rate of each fungal strain to be employed. As for *T. virens*, 40 h is chosen because its colony reaches the edge of the membrane (~8 mm away). Once fungal mycelia grow beyond the membrane and contaminate the medium, the plate cannot be used to assay the effect of *F. oxysporum* VCs on the secretion of antifungal molecules. Check the growth rate of tested strains on a cellophane membrane before setting up large-scale experiments.
11. The harvested *Trichoderma* culture, along with the cellophane membrane, should be immediately frozen in liquid nitrogen and stored at −80 °C for subsequent analyses (e.g., transcripts, proteins, metabolites). Flame sterilize the forceps between plates to prevent contamination.
12. Hyphae of some *F. oxysporum* isolates can breach the membrane, presumably by secreting cellulolytic enzymes that degrade the membrane, and contaminate the medium during treatment. Check the ability of tested strains to degrade the cellophane membrane before setting up large-scale experiments.
13. In the study by Li et al. [27], we included *Escherichia coli* BW25113, a strain used for producing the Keio collection of genome-wide mutants [36]. This strain was evaluated along with bacteria isolated from the rhizosphere of tomato to determine if its mutant collection can be screened to identify the genes involved in producing VCs that affect *F. oxysporum*. Culture conditions and the duration of the treatment for individual bacterial species need to be evaluated before setting up the treatments noted here.
14. Too much water on the surface prevents the formation of distinct bacterial colonies and causes streaks of cells, which prevents the quantification of the VC-mediated effect on bacterial growth.
15. Besides counting the number of CFUs, visually compare the size of colonies between the treatment and control as an indicator of bacteriostatic activity.

16. This option can be employed when the antibacterial activity of molecules secreted by the tested fungal strain is so high that the bacterial cells cannot grow on the treated plates.
17. Different coating materials for the SPME fiber may display different affinities to identical compounds, potentially affecting the composition of captured chemical profiles depending on which fiber has been used. Compounds with high affinity may displace previously bound VCs with lower affinity from the fiber; the duration of VC collection also affects the resulting profiles, contributing to sampling biases. Different absorbents (e.g., SuperQ) can be used [22].
18. The optimal duration of cultivation will need to be evaluated for individual strains and species. For some experiments, sampling at multiple time points may be necessary.

Acknowledgments

The protocols and systems described here have been developed using support from the USDA Specialty Crop Multi-State Program (AM170200XXXXG006), the USDA National Institute of Food and Agriculture and Federal Appropriations (Project PEN04655; Accession # 1016291), the Chinese Academy of Sciences President's International Fellowship Initiative (Grant # 2019VCA0022), and the Brian Pool program of the National Research Foundation of Korea (Grant # 2019H1D3A2A01054562).

References

1. Kang S, Demers J, del Mar Jimenez-Gasco M et al (2014) Fusarium oxysporum. In: Dean RA, Lichens-Park A, Kole C (eds) Genomics of plant-associated fungi and oomycetes: dicot pathogens. Springer, Berlin, pp 99–119
2. Gordon TR, Martyn RD (1997) The evolutionary biology of *Fusarium oxysporum*. Annu Rev Phytopathol 35:111–128
3. Benhamou N, Garand C, Goulet A (2002) Ability of nonpathogenic *Fusarium oxysporum* strain Fo47 to induce resistance against *Pythium ultimum* infection in cucumber. Appl Environ Microbiol 68:4044–4060
4. Nucci M, Anaissie E (2007) Fusarium infections in immunocompromised patients. Clin Microbiol Rev 20:695–704
5. Husaini AM, Sakina A, Cambay SR (2018) Host–pathogen interaction in *Fusarium oxysporum* infections: where do we stand? Mol Plant-Microbe Interact 31:889–898
6. del Rosario Cappellari L, Chiappero J, Banchio E (2019) Invisible signals from the underground: a practical method to investigate the effect of microbial volatile organic compounds emitted by rhizobacteria on plant growth. Biochem Mol Biol Educ 47:388–393
7. Kesselmeier J, Staudt M (1999) Biogenic volatile organic compounds (VOC): an overview on emission, physiology and ecology. J Atmos Chem 33:23–88
8. Schulz-Bohm K, Martín-Sánchez L, Garbeva P (2017) Microbial volatiles: small molecules with an important role in intra- and inter-kingdom interactions. Front Microbiol 8:2484
9. Schulz-Bohm K, Gerards S, Hundscheid M et al (2018) Calling from distance: attraction of soil bacteria by plant root volatiles. ISME J 12:1252–1262
10. Erb M (2018) Volatiles as inducers and suppressors of plant defense and immunity —

origins, specificity, perception and signaling. Curr Opin Plant Biol 44:117–121

11. Das A, Lee SH, Hyun TK et al (2013) Plant volatiles as method of communication. Plant Biotechnol Rep 7:9–26
12. Kant MR, Bleeker PM, Wijk MVAN et al (2009) Plant volatiles in defence. Adv Bot Res 51:613–666
13. Dudareva N, Negre F, Nagegowda DA et al (2006) Plant volatiles: recent advances and future perspectives. Crit Rev Plant Sci 25:417–440
14. Law JH, Regnier FE (1971) Pheromones. Annu Rev Biochem 40:533–548
15. Bitas V, Kim H-S, Bennett JW et al (2013) Sniffing on microbes: diverse roles of microbial volatile organic compounds in plant health. Mol Plant-Microbe Interact 26:835–843
16. Effmert U, Kalderás J, Warnke R et al (2012) Volatile mediated interactions between bacteria and fungi in the soil. J Chem Ecol 38:665–703
17. Audrain B, Farag MA, Ryu C-M et al (2015) Role of bacterial volatile compounds in bacterial biology. FEMS Microbiol Rev 39:222–233
18. Chernin L, Toklikishvili N, Ovadis M et al (2011) Quorum-sensing quenching by rhizobacterial volatiles. Environ Microbiol Rep 3:698–704
19. Li N, Alfiky A, Vaughan MM et al (2016) Stop and smell the fungi: fungal volatile metabolites are overlooked signals involved in fungal interaction with plants. Fungal Biol Rev 30:134–144
20. Piechulla B, Lemfack MC, Kai M (2017) Effects of discrete bioactive microbial volatiles on plants and fungi. Plant Cell Environ 40:2042–2067
21. Sharifi R, Ryu CM (2018) Sniffing bacterial volatile compounds for healthier plants. Curr Opin Plant Biol 44:88–97
22. Bitas V, McCartney N, Li N et al (2015) Fusarium oxysporum volatiles enhance plant growth via affecting auxin transport and signaling. Front Microbiol 6:1248
23. Minerdi D, Bossi S, Gullino ML et al (2009) Volatile organic compounds: a potential direct long-distance mechanism for antagonistic action of *Fusarium oxysporum* strain MSA 35. Environ Microbiol 11:844–854
24. Minerdi D, Bossi S, Maffei ME et al (2011) Fusarium oxysporum and its bacterial consortium promote lettuce growth and expansin A5 gene expression through microbial volatile organic compound (MVOC) emission. FEMS Microbiol Ecol 76:342–351
25. Li N, Alfiky A, Wang W et al (2018) Volatile compound-mediated recognition and inhibition between *Trichoderma* biocontrol agents and *Fusarium oxysporum*. Front Microbiol 9:1–16
26. Medina-Romero YM, Roque-Flores G, Macías-Rubalcava ML (2017) Volatile organic compounds from endophytic fungi as innovative postharvest control of *Fusarium oxysporum* in cherry tomato fruits. Appl Microbiol Biotechnol 101:8209–8222
27. Li N, Islam MT, Kang S (2019) Secreted metabolite-mediated interactions between rhizosphere bacteria and *Trichoderma* biocontrol agents. PLoS One 14:e0227228
28. Li N, Kang S (2018) Do volatile compounds produced by *Fusarium oxysporum* and *Verticillium dahliae* affect stress tolerance in plants? Mycology 69:166–175
29. Hoagland DR, Arnon DI (1950) The water-culture method for growing plants without soil. Calif Agric Exp Stn Circ 347:1–32
30. Kai M, Piechulla B (2009) Plant growth promotion due to rhizobacterial volatiles - an effect of CO_2? FEBS Lett 583:3473–3477
31. Wenke K, Wanke D, Kilian J et al (2012) Volatiles of two growth-inhibiting rhizobacteria commonly engage AtWRKY18 function. Plant J 70:445–459
32. Kwon YS, Ryu C-M, Lee S et al (2010) Proteome analysis of *Arabidopsis* seedlings exposed to bacterial volatiles. Planta 232:1355–1370
33. Hiscox JD, Israelstam GF (1979) A method for the extraction of chlorophyll from leaf tissue without maceration. Can J Bot 57:1332–1334
34. Wang W, Li N, Liu X et al (2019) I plate-based assay for studying how fungal volatile compounds (VCs) affect plant growth and development and the identification of VCs via SPME-GC-MS. Bio-Protocol 9:e3166
35. Kellogg J, Kang S (2020) Metabolomics, an essential tool in exploring and harnessing microbial chemical ecology. Phytobiomes J 4:195–210
36. Baba T, Ara T, Hasegawa M et al (2006) Construction of *Escherichia coli* K-12 in-frame, single-gene knockout mutants: the Keio collection. Mol Syst Biol 2:2006.0008

Chapter 11

Isolation of Mitochondria from *Fusarium* Species

Michael C. MacKillop and John C. Kennell

Abstract

Isolation of purified mitochondria is an essential technique for the analysis of metabolic and cellular functions associated with this vital organelle. Filamentous fungi, such as *Neurospora crassa*, have been proven to be highly amenable to the analysis of mitochondria, in part due to their rapid growth rate and relative ease of isolation. Here we describe a step-by-step procedure for the isolation of mitochondria from *Fusarium* species via differential centrifugation and density step-gradient centrifugation, and include methods to overcome potential complications. Mitochondria purified by flotation gradient procedures remain active for functional assays and can be further fractionated for isolation of nucleic acids or ribonucleoprotein particles that retain enzymatic activity.

Key words Cellular fractionation, Isolation and purification of mitochondria, Flotation gradient

1 Introduction

Isolation of mitochondria is an important technique for mitochondrial studies, and, due to the amenability of genetic, biochemical, and molecular analyses studies of yeasts and filamentous fungi have provided fundamental insights into mitochondrial biology. This chapter focuses on procedures for the purification of mitochondria from *Fusarium* species that can be used in functional assays, as well as the isolation of mitochondrial nucleic acids and ribonucleoprotein complexes that retain enzymatic activity.

George Palade and coworkers pioneered the isolation of mitochondria from animal liver cells by procedures involving differential centrifugation [1]. In the 1970s, Luck and coworkers developed a flotation gradient method for the isolation of mitochondria and mitochondrial ribosomes, free from contaminating cytosolic ribosomes, from *Neurospora crassa* [2, 3]. Additional protocols have been since described, such as the use of Percoll for subfractionation [4]. When used in combination, the "quick and dirty" steps of differential centrifugation and flotation gradient centrifugation provide an effective procedure that can be scaled up to isolate

Jeffrey Coleman (ed.), *Fusarium wilt: Methods and Protocols*,
Methods in Molecular Biology, vol. 2391, https://doi.org/10.1007/978-1-0716-1795-3_11,

gram-quantities of purified mitochondria. In most instances, the costs can be kept to a minimum by the use of sucrose gradients rather than Percoll without compromising the quality of the preparations. Isolated mitochondria can, in turn, be used to study fundamental processes including mitochondrial respiration, metabolism, polymerase and nuclease activities, protein import, protein complex assembly, mitochondrial protein synthesis, as well as "omic" analyses and interactions of mitochondria with other organelles and cytosolic compounds.

The techniques described in this chapter derive from studies with the filamentous fungus, *Neurospora crassa*. *Neurospora* has a rich history as a model system for mitochondrial studies, in part due to the rapid vegetative growth and relative ease of isolation of mitochondria from mycelia. The isolation of mitochondria from *Fusarium* species is complicated by the production of macro- and micro-conidia that make it challenging to accurately determine inoculum levels, and the production of hydrophobins by many species [5, 6] that interfere with isolation. Steps are described to minimize these complications.

2 Materials

Growth

1. 50 mL of Vogel's media [7, 8] in 250 mL Erlenmeyer flasks with aeratable stopper (*see* **Note 1**). Vogel's Medium N is made by preparing a 1 L 50-strength solution (50×) that can be stored at room temperature indefinitely. Begin with 750 mL of deionized water, and dissolve the ingredients in the order below, allowing each to be fully dissolved before adding the next ingredient: 150 g Na_3-Citrate·5 H_20; 250 g KH_2PO_4; 100 g NH_4NO_3; 10 g $MgSO_4$·7 H_2O; 5 g $CaCl_2$·2 H_2O (*see* **Note 2**); 5 mL of fresh biotin solution [5 mg of biotin in 100 mL of 50% (v/v) ethanol, stored at −20 °C]; 5 mL of trace element solution [a 100 mL stock is prepared by adding the following ingredients in order to 95 mL of deionized water: 5 g Citric acid · H_2O; 5 g $ZnSO_4$ · 7 H_2O; 1 g Fe $(NH_4)_2(SO_4)_2$ · 6 H_2O; 250 mg $CuSO_4$ · 5 H_2O; 50 mg $MnSO_4$ · H_2O; 50 mg H_3BO_3; 50 mg Na_2MoO_4 · H_2O, final volume adjusted to 100 mL] . The 50× stock should be brought to a final volume of 1 L and 2–3 mL of chloroform is added as a preservative. A single-strength (1×) solution is made by adding 1 part stock with 49 parts deionized water, adding sucrose to 1.5%, and autoclaving.
2. 1 L of Vogel's media in a 4 L flask with an aeratable stopper.
3. 15 mL or 50 mL sterile conical tubes.

4. Sterile 100% glycerol.
5. Hemocytometer.
6. Microscope with 10× and 40× objective lenses.
7. Floor shaker or shaker-incubator with capacity to shake 250 mL and 4 L flasks.

Differential Centrifugation

1. Buchner funnels (2 L capacity) (*see* **Note 3**).
2. Filter Paper (Grade #417, qualitative).
3. Bead Beater (250 mL capacity).
4. Beads (0.5 mm glass beads).
5. 10 mL serological pipets (standard taper and wide mouth) and pipet pumps.
6. 15% Sucrose solution: In acid-cleaned glassware, make 3 L of the solution composed of 10 mM Tricine-KOH, 0.2 mM EDTA, pH 7.5, and 15% sucrose. Store in a sterilized container at 4 °C.
7. 20% Sucrose solution: In acid-cleaned glassware, make 1 L of the solution composed of 10 mM Tricine-KOH, 0.2 mM EDTA, pH 7.5, and 20% sucrose. Store in a sterilized container at 4 °C.

Flotation Gradient

1. 44% Sucrose solution: In acid-cleaned glassware, prepare 1 L of 10 mM Tricine-KOH, pH 7.5, and 44% ultra-pure sucrose. Store in a sterilized container at 4 °C.
2. 53% Sucrose solution: In acid-cleaned glassware, prepare 1 L of 10 mM Tricine-KOH, pH 7.5, and 53% ultra-pure sucrose. Store in sterilized container at 4 °C.
3. 70% Sucrose solution: In acid-cleaned glassware, prepare 1 L of 10 mM Tricine-KOH, pH 7.5, and 70% ultra-pure sucrose. Store in sterilized container at 4 °C.
4. HKCT+D solution: 500 mM KCl, 25 mM $CaCl_2$, 25 mM Tris–HCl, pH 7.5. Fresh DTT should be added prior to use. Add 0.1155 g DTT per 150 mL of HKCT. Store in sterilized container at 4 °C.
5. High Speed centrifuge.
6. Fixed angle rotor (e.g., GSA).
7. 12–250 mL centrifuge tubes.
8. Fixed angel rotor (e.g., Sorvall SS34).
9. 12–50 mL centrifuge tubes.
10. Ultracentrifuge.

11. SW-28 swinging bucket rotor or equivalent.
12. 6 SW-28 centrifuge tubes.

3 Methods

3.1 Conidial Stocks

1. Inoculate small volumes (10–50 mL) of Vogel's media in 250 mL flasks with conidia isolated from a plate or directly inoculated from a frozen stock.
2. Shake flasks at 200 rpm in a shaker-incubator or free-standing shaker at room temperature for 24–48 h until cultures reach a high density.
3. Aseptically remove an aliquot of the fungal culture to observe under the microscope, and check for contamination. Estimate and record the relative percentage of macroconidia, microconidia, germinating conidia, and broken hyphal cells (*see* **Note 4**).
4. Add 3 mL of 100% sterile glycerol to a 15 mL sterile conical tube (or 10 mL to a 50 mL conical tube), and add conidial solution to top and invert before labeling and storing in a −80 °C freezer (*see* **Note 5**).

3.2 Vegetative Growth

1. In a laminar flow biosafety cabinet, inoculate 4 L flasks containing 1 L of Vogel's media to a final concentration of approximately 1×10^5 cells/mL (*see* **Note 6**).
2. Shake the flasks at 200 rpm for 24–36 h at room temperature until there are traces of mycelia clinging to the sides of the flask.
3. Turn off shaker and allow conidia to germinate for 3–7 days, until a mycelial mat has formed on the top of the media. Periodically (once a day), flasks can be swirled to remove mycelia that form on the sides of the flasks (*see* **Note 7**).

3.3 Preparation of Fusarium Mitochondria

All steps should be performed at 4 °C or on ice unless otherwise indicated. Treat any spillage on counters with freshly prepared 10% bleach followed by 70% ethanol.

1. Harvest mycelia from 4 L flasks by pouring through the Buchner funnel containing a moistened sheet of #417 filter paper (*see* **Note 8**). Scrape pads from filter paper, combine, and save in foil on ice. Record the weight.
2. Fill bead beater container a third of the way with glass beads, followed by mycelia torn into small pieces to cover the beads, and add the 15% sucrose solution until the container is filled close to the top. Screw on lid, and check for leaks. Put onto bead beater and use 2 or 3 short "blasts" to create homogenous slurry. Beat solution 3 times for 10 s, allowing beads to settle

for 15 s between runs. Decant supernatant to a sterile 2 L flask, being careful not to transfer beads. Repeat until all mycelial is ground.

3. Stir and decant equal amounts of the lysate into 250 mL centrifuge bottles. Add additional 15% sucrose solution, and repeat until the beads are clean or until GSA bottles are full (*see* **Note 9**).
4. Spin at 2600 × *g* for 10 min 4 °C (*see* **Note 10**).
5. Pour the supernatant into fresh 250 mL centrifuge bottles, balance by visual inspection, and spin at 18,600 × *g* for 25 min 4 °C.
6. Decant supernatant, add 5 mL of the 20% sucrose solution to the pellets, and resuspend carefully with a wide mouth serological pipette. Transfer the 20% sucrose solution containing the mitochondria to a 50 mL centrifuge tube. Rinse the 250 mL tube with 5 mL of the 20% sucrose solution, and combine for a total of 10 mL. Centrifuge at 26,900 × *g* for 20 min. Decant supernatant.

   Stopping Point (*see* Fig. 1). This pellet can be stored at −80 °C until ready to proceed with flotation gradient purification (*see* **Note 11**).
7. Using a serological pipette, resuspend the pellet with 5 mL of the 70% sucrose solution. Transfer the 5 mL suspension to a SW28 tube. Rinse the 50 mL tube with 10 mL of the 70% sucrose solution, and add to the 5 mL in the SW28 tube for a total of 15 mL and mix thoroughly (*see* **Note 12**).
8. Overlay with 12 mL of the 53% sucrose solution, followed by 10 mL of the 44% sucrose solution (Fig. 2a) (*see* **Note 13**).
9. Balance the tubes by weighing and adding more of the 44% sucrose solution to top if necessary. Place the sucrose gradient tubes into the centrifuge buckets, and carefully hang the buckets on the rotor. Place into the ultracentrifuge and run with a vacuum at 131,000 × *g*, at 4 °C, for 140 min.
10. Following a slow deceleration, remove the rotor and place the buckets on ice. Carefully remove the tubes with forceps and gently place in ice (Fig. 2b). Use a 10 mL pipet (standard or narrow taper) to remove ~10–15 mL of the mitochondria that localize at the interface of 53/44% sucrose layer, and place into a clean 50 mL tube. Add at least 2.5–3 volumes of HKCT+D solution to the mitochondrial sucrose solution, and invert tube to mix. Centrifuge at 21,000 × *g*, 4 °C for 10 min (*see* **Note 14**).
11. Aspirate off the supernatant (Fig. 2c). Label the tube and place in a −80 °C freezer (*see* **Note 15**).

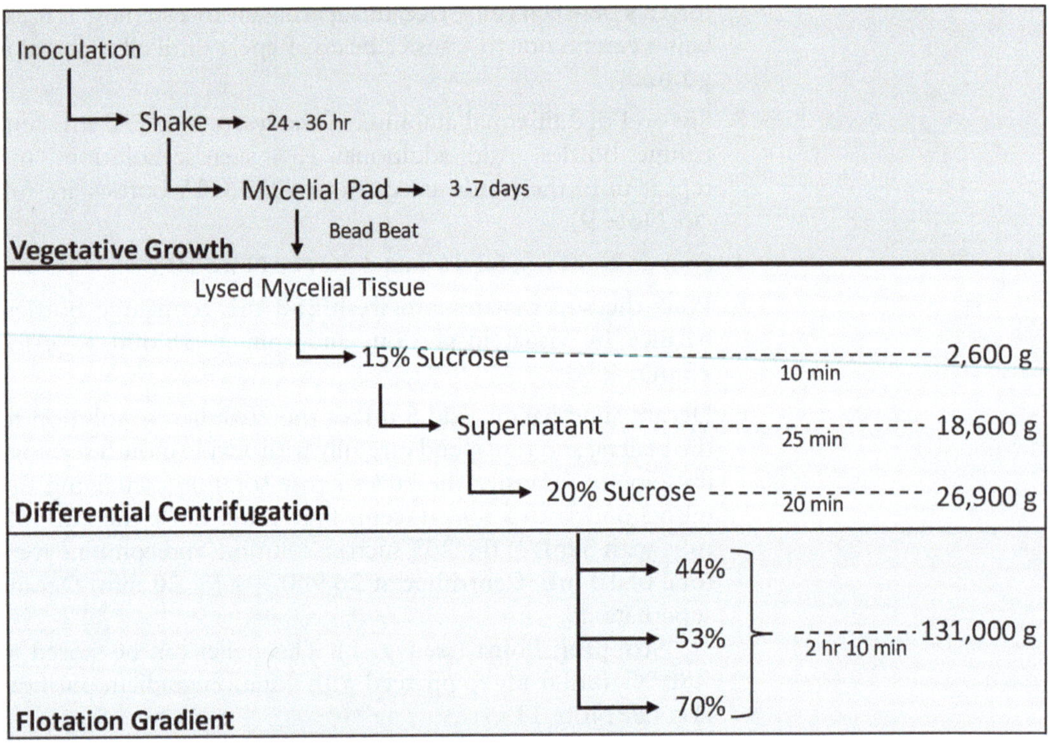

Fig. 1 Workflow overview for purification of *Fusarium* mitochondria

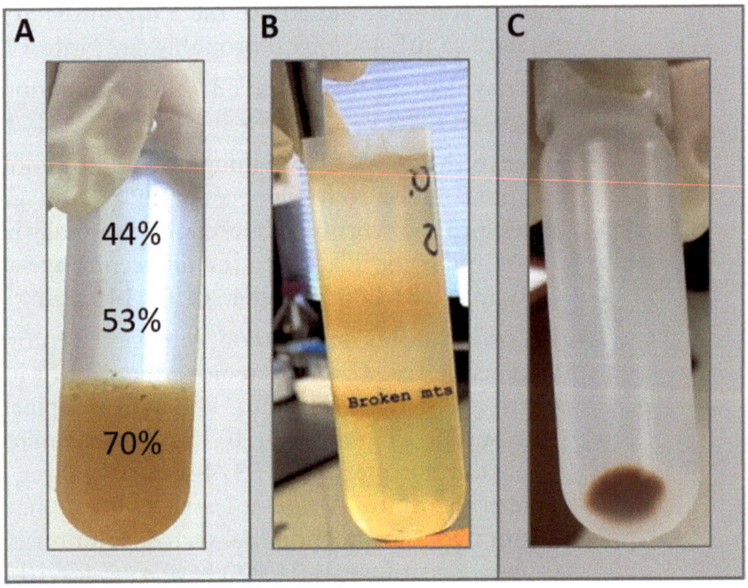

Fig. 2 Purified mitochondria from flotation gradient centrifugation. (**a**) Crude mitochondrial extract resuspended in 15 mL 70% sucrose solution, layered with 15 mL of 53% sucrose solution and 10 mL of 44% sucrose solution. (**b**) Flotation gradient after 2 h at 131,000 × *g*. No more than 10 mL of purified mitochondria should be pipetted from the 44/53% sucrose interface. (**c**) Final mitochondrial pellet after 21,000 × *g* spin in 2–3× HKCT+D

4 Notes

1. Growth conditions that favor hyphal growth (rather than conidiation) are important for successful isolation of mitochondria. Growth media used may vary for particular taxa. For *Fusarium* species or strains that are to be repeatedly studied, conditions favorable for both conidial and mycelial production should be empirically determined. The procedures described below utilize Vogel's Medium N [7], but can also be used with other types of growth media, such as potato dextrose media [8].
2. Pre-dissolving $CaCl_2$ in 20 mL of H_2O helps to prevent precipitation.
3. Procedure can be scaled down with appropriate fixed/swinging bucket rotors and centrifuge tubes.
4. For fungal species such as *N. crassa* that readily produce homogeneous macroconidia, the conidial concentration is calculated using a hemocytometer. The target concentration for a concentrated stock solution is $>7.5 \times 10^5$ condia/μL. For *Fusarium* species, conidial types are typically mixed with hyphae and difficult to accurately quantitate. Instead, cell types (e.g., macroconidia, microconidia, germinating conidia, broken hyphal cells) are recorded, and the density is estimated for total cells (optimally, in the range of 1×10^5 cells/μL). If a strain is to be routinely used for analysis, multiple conidial cultures are grown and combined, prior to storage in 15 or 50 mL sterile conical tubes.
5. To avoid freeze-thawing cultures, aliquots of the conidia can also be stored in 1.5 mL sterile microcentrifuge tubes.
6. A preparation using 8 flasks will typically produce greater than 250 g of mycelial mass that will yield 4–6 nickel-sized mitochondrial pellets following the flotation density isolation method. There can be large differences in the yield depending on the species or strain with the variation most directly associated with degree of vegetative growth. In our experience, *F. oxysporum* strains are more amenable to producing mycelial mats and have less hydrophobin production that interfere with fractionation of mitochondria in differential centrifugation and flotation gradient steps.
7. The length of incubations can vary, depending on the strain, growth media, and inoculation density.
8. A trap can be installed by inserting rubber plug into top of collecting flask, and attaching a HEPA filtered tubing to the rubber plug on one end, and vacuum filtrate station on the other end.

9. A typical prep will yield ~1500 mL of the mycelial/sucrose slurry. Significant insoluble material in the supernatant after the 2600 × *g* spin indicates the slurry was too thick and should be diluted. Multiple rounds of centrifugation can be performed as necessary based on the starting amount of mycelial mass.
10. Save the material pelleted from this step. If a dark-brown pellet is not observed after the centrifugation in the next step, then this pellet can be re-extracted with fresh sucrose, following the procedure of **step 2**.
11. The pellet from **step 6** contains "crude" or partially purified mitochondria. This pellet can be used for the isolation of mitochondrial DNA, though significant cytosolic impurities will be present.
12. If the solution contains aggregated material, drawing the solution in and out of a 1 mL or 5 mL disposable pipette tip can be used to break up the clumps.
13. Care should be taken when layering the sucrose solutions. Proper layering can be achieved by slowly dispensing the sucrose solution in a circular motion around centrifuge tube walls.
14. Use a sterile cotton swab to remove any cellular debris or hydrophobins from top of 44% layer before extracting the purified mitochondria. Using a volume less than 2.5× of HKCT+ D will prevent mitochondria from precipitating in the subsequent centrifugation step due to the high concentration of sucrose.
15. If subsequent analyses can be carried out with small quantities of mitochondria, the mitochondrial pellets can be resuspended in HKCT+ D (or other buffers) and aliquoted into microcentrifugation tubes, pelleted, and stored at −80 °C. The quantity of mitochondria can be estimated by measurement of proteins by the Biuret or Bradford methods. Alternatively, the mitochondrial pellets can be resuspended directly for isolation of mitochondrial ribonucleoprotein particles that retain enzymatic activity [9].

References

1. Hogeboom GH, Schneider WC, Pallade GE (1948) Cytochemical studies of mammalian tissues. I. Isolation of intact mitochondria from rat liver; some biochemical properties of mitochondria and submicroscopic particulate material. J Biol Chem 172:619–635
2. Lizardi PM, Luck DJL (1971) Absence of a 5S RNA component in the mitochondrial ribosomes of *Neurospora crassa*. Nat New Biol 229:140
3. Lambowitz AM, Chua NH, Luck DJ (1976) Mitochondrial ribosome assembly in *Neurospora*. Preparation of mitochondrial ribosomal precursor particles, site of synthesis of mitochondrial ribosomal proteins and studies on the poky mutant. J Mol Biol 107:223–253

4. Pertoft H (2000) Fractionation of cells and subcellular particles with Percoll. J Biochem Biophys Methods 44:1–30
5. Wösten H (2001) Hydrophobins: multipurpose proteins. Annu Rev Microbiol 55:625–646
6. Quarantin A, Hadeler B, Kröger C et al (2019) Different hydrophobins of *Fusarium graminearum* are involved in hyphal growth, attachment, water-air interface penetration and plant infection. Front Microbiol 10:751
7. Vogel HJ (1964) Distribution of lysine pathways among fungi: evolutionary implications. Naturalist 98:435
8. Davis RH, de Serres FJ (1970) Genetic and Microbiological Research Techniques for *Neurospora crassa*. Methods Enzymol 17:79–143
9. Walther TC, Kennell JC (1999) Linear mitochondrial plasmids of *F. oxysporum* are novel, telomere-like retroelements. Mol Cell 4:229–238

Chapter 12

In Planta Gene Expression Analysis and Colonization of *Fusarium oxysporum*

Virginia Casado-del Castillo, Ernesto Pérez Benito, and José María Díaz-Mínguez

Abstract

In planta gene expression analysis and GFP-based confocal microscopy are two powerful techniques that may be coupled to assess the extent and dynamics of plant colonization by a fungal pathogen. Here we describe methods to prepare common bean plants for inoculation with a highly virulent strain of *Fusarium oxysporum* f. sp. *phaseoli*, quantify the extent of colonization by RT-qPCR, and visualize the colonized tissues by confocal microscopy.

Key words *Fusarium oxysporum*, *Phaseolus vulgaris*, RT-qPCR, Confocal laser microscopy, Hydroponic cultures

1 Introduction

The study of genes involved in pathogenicity and their evolution has attracted considerably more attention by plant pathologists than the study of virulence and how virulence genes have evolved. It is clear that understanding the evolution of pathogenicity is of paramount importance for the control of plant diseases, as it determines the success and durability of resistant cultivars. However, the evolution of virulence is fundamental to understanding the life history of pathogens. We understand virulence as the degree of damage caused to a host by parasite infection, while pathogenicity is the qualitative capacity of a parasite to infect and cause disease on a host [1]. The preference in the analysis of pathogenicity over virulence may be explained by the nature of the tools available. A favorite strategy to shed light on the genetic basis of pathogenicity is gene deletion. By using plant mutants, inactivated in a resistance gene, and parasite mutants, inactivated in a pathogenicity gene, and verifying the outcome of the interaction by means of an infection assay, it is possible to determine gene-for-gene interactions in host-

Jeffrey Coleman (ed.), *Fusarium wilt: Methods and Protocols*,
Methods in Molecular Biology, vol. 2391, https://doi.org/10.1007/978-1-0716-1795-3_12,

parasite systems. The rationale behind this analysis is the qualitative effect of resistance and pathogenicity genes, which usually determines clear results that are easy to interpret.

It is more difficult to use this strategy to determine the role of virulence genes, as its inactivation produces quantitative effects which more often are not easily shown in the infection assays. Pathogenicity assays may discriminate strains which are less or more aggressive [2], but are not very useful to quantify differences in virulence caused by the expression of individual genes, unless they behave functionally as major virulence factors.

Transmission electron microscopy was used in the pioneering studies carried out by Bishop and Cooper to unravel the differences in root invasion by vascular pathogens when confronting resistant and susceptible cultivars of tomato and the characteristics of vascular colonization of tomato by different vascular pathogens [3], including *F. oxysporum*. A major breakout in the visualization of vascular colonization was the development of laser scanning spectral confocal microscopy (CLSM) and the use of several variants of the green fluorescent protein (GFP). This technology allows in vivo and time-lapse imaging of plant pathogenesis, avoiding problems associated with fixing plant tissues for light and electron microscopy observations [4]. Confocal laser microscopy has allowed for a better understanding of the infection process of *F. oxysporum* in multiple hosts and different pathogenic races in compatible and incompatible interactions [5, 6]. Also, and of not lesser importance, it may be used to study the patterns of colonization displayed by strains showing differences in virulence [7]. For example, it has helped to demonstrate how the speed of xylem vessel colonization determines disease severity [8].

RT-qPCR is the best suited method for high-sensitivity quantification of gene expression. In addition, RT-qPCR assays can detect small differences in gene expression, can also discriminate between mRNAs with very similar sequences, and requires much less RNA template than other methods of gene expression analysis. These advantages make RT-qPCR the method of choice for the analysis of variable transcript levels in host plant tissues during fungal pathogenesis. When used in combination with confocal laser microscopy, the effect of transient expression of virulence genes may be correlated with the colonization pattern displayed inside the host by the pathogen, allowing for an in-depth functional analysis of the role of virulence genes during infection [7]. The development of non-specific dsDNA binding dyes, such as SYBR Green, has enabled the development of rapid and cheap assays well suited for the analysis of temporal expression of multiple genes during host colonization by fungal pathogens. However, the extremely high sensitivity makes it essential to apply well-designed normalization techniques, especially when the expression of several genes is quantified during different life stages of the host and the pathogen [9].

2 Materials

2.1 Plant Inoculation with *Fusarium oxysporum*

2.1.1 Growth of Plants

1. Vermiculite (Projar, SA): sterilized in an autoclave, two cycles of 20 min at 121 °C each.
2. Commercial bleach 5% (prepared in distilled water).

2.1.2 Preparation of Conidia to Be Used as Inoculum

1. Sporulation medium [10] (per liter of distilled water): 60 g sucrose, 7 g $NaNO_3$, 3 g tryptone, 1 g KH_2PO_4, 0.5 g $MgSO_4 \cdot 7\ H_2O$, 0.5 g KCl, pH 7.0.
2. Sterilized funnel with double layer of cheese cloth.
3. Centrifuge.
4. Sterile distilled water (SDW).
5. Hemocytometer.

2.1.3 Inoculation of *Phaseolus vulgaris*

1. Plant Growth Medium (PGM solution) [11] (per liter of distilled water): 0.24 g $MgSO_4 \cdot 7\ H_2O$, 0.04 g K_2HPO_4, 0.17 g K_2SO_4, 0.344 g $CaSO_4 \cdot H_2O$, 0.64 g NH_4NO_3, 1 mL trace elements (*see* **Note 1**).
2. Trace elements solution [11] (per liter of distilled water): 0.375 g $ZnSO_4 \cdot 7\ H_2O$, 0.125 g $CuSO_4 \cdot 5\ H_2O$, 0.846 g $MnSO_4 \cdot H_2O$, 1.546 g H_3BO_3, 0.05 g $FeCl_3$, 0.728 g KCl, 0.081 g H_2MoO_4, 0.001 g $CoCl_2 \cdot 6\ H_2O$ pH 7.
3. Sterile 50 mL centrifuge tubes.
4. Parafilm.
5. Aluminum foil.
6. Laminar-flow hood.
7. Controlled environment chamber.

2.2 RT-qPCR Quantification and Analysis

2.2.1 Isolation of RNA

1. SV Total RNA Isolation System (Promega).
2. DNase: Turbo DNA-*free*™ kit (Ambion).

2.2.2 Determination of RNA Concentration and Quality

1. 5× MOPS Buffer (prepared with DEPC-treated water): 0.02 M 3-morpholinopropanesulfonic acid (MOPS) (pH 7.0), 0.05 M sodium acetate, 0.005 M EDTA (pH 8.0). Sterilize in an autoclave (*see* **Note 2**).
2. RNA Sample Buffer: 10 mL deionized formamide, 3.5 mL 37% formaldehyde, 5× MOPS Buffer. Dispense into single use aliquots and store at −20 °C in tightly sealed tubes (*see* **Note 3**).

3. RNA Loading Buffer (prepared with DEPC-treated water): 50% glycerol, 1 mM EDTA (pH 8.0), 0.4% bromophenol blue. Dispense into single-use aliquots and store at −20 °C (*see* **Note 4**).
4. 37% formaldehyde.
5. Nanodrop spectrophotometer.
6. Agarose.
7. Hotplate.

2.2.3 cDNA Synthesis

1. PrimeScript™ RT Reagent kit (Perfect Real Time, Takara Biosystems).
2. Thermocycler.

2.2.4 RT-qPCR

1. Sterile 96-well plate. Select the appropriate plate according to your thermocycler.
2. Polyester heat seal: ThermaSeal ATM, 50 μm thickness, 41 μm adhesive layer (CMB Bioline).
3. KAPA™ SYBR® FAST qPCR Kit Master Mix (2×) ABI Prism™.
4. Specific primer pairs for each gene to be analyzed.
5. Thermocycler for real-time qPCR (we use the Step One Plus™ Real-Time PCR System, Applied Biosystems).

2.3 Confocal Laser Microscopy

1. PBS (Phosphate Buffered Saline) 1× (per liter of distilled water): 8.06 g NaCl, 0.22 g KCl, 1.15 g Na_2HPO_4, 0.20 g KH_2PO_4, pH 7.4.
2. Propidium iodide stock solution (prepared in PBS): 1 mg/mL propidium iodide, pH 7.4. Store at 4 °C in the dark.
3. 10% KOH (prepared in distilled water).
4. WGA Alexa Fluor 488™ (Thermo Fisher Scientific) stock solution: 1 mg/mL of sterile distilled water. Store at −20 °C in the dark.
5. Staining Solution (prepared in PBS 1×): 10 μg/mL WGA (10 μL/mL of WGA stock solution), 20 μg/mL propidium iodide (20 μL/mL of propidium iodide stock solution), 0.02% (V/V) Tween 20.
6. Glycerol 30% (prepared in distilled water).
7. Centrifugal vacuum concentration system.
8. Laser confocal microscope (we use TCS2-SP2, Leica Microsystems, Germany).
9. 100% ethanol.

3 Methods

3.1 Plant Inoculation with F. oxysporum

Phaseolus vulgaris L. plants inoculated with conidia of *F. oxysporum* were grown as hydroponic cultures in an environment-controlled chamber. This method allows for the collection of clean samples well suited for RT-qPCR and microscopic analysis.

3.1.1 Growth of Plants

1. Surface sterilize bean seeds by washing them with 5% commercial bleach for 5 min.
2. Remove the bleach by washing the seeds with distilled water 4–6 times for 2 min each time.
3. Dry the sterilized seeds and place them in vermiculite for germination. Incubate in an environment-controlled chamber or in greenhouse at 23–25 °C with a 12-h photoperiod and relative humidity of 60–80% (Fig. 1a).

3.1.2 Preparation of the Inoculum

1. Grow the fungal strain(s) on a shaker in 50 mL of sporulation medium for 4–5 days at 25 °C and 180 rpm with continuous light.
2. Harvest the conidia the same day the plants will be inoculated. Filter the culture from **step 1** through a double layer of sterile cloth, and recover the conidia, mostly microconidia, by centrifugation at 2600 × *g* for 5 min.
3. Wash the microconidia with sterile distilled water (SDW) 3–4 times, and collect them by centrifugation as described in **step 2**.
4. Use a hemocytometer to count the number of microconidia, and adjust the conidial suspension to 10^6 conidia per mL of SDW. The final volume of the conidial suspension must be enough to completely cover the roots of the plants to be inoculated.

Fig. 1 Inoculation of common bean plants by means of a standard root-dip technique and grown as hydroponic cultures. (**a**) 6–8 days *P. vulgaris* plantlets germinated in vermiculite. (**b–c**) Inoculated common bean plant 5 days post infection (dpi) growing in PGM solution (hydroponic culture)

3.1.3 Inoculation of *Phaseolus vulgaris*

The inoculation of *P. vulgaris* with *F. oxysporum* follows a standard root-dip technique [12–14], which is highly effective to induce *Fusarium* wilt in a homogeneous way, and is easy and quick to carry out.

1. Prepare 8–10-day-old seedlings grown in vermiculite (the unifoliate leaves must be at three-quarters of full expansion).
2. Wash the roots in running tap water to remove the vermiculite.
3. Cut off about 1 cm from the tip of the roots.
4. Inoculate the plants by dipping the roots for 5 min in the previously prepared conidial suspension. For mock inoculations, dip the roots in SDW.
5. After the inoculation, transfer the plants to sterile 50 mL centrifuge tubes filled with PGM solution, seal the top of the tube with Parafilm®, and cover the tube with foil. This step must be carried out inside a laminar-flow hood to avoid contamination of the PGM solution (Fig. 1b, c).
6. Incubate the inoculated plants in a controlled environment chamber at 25 °C and relative humidity of 60–80% with a 16 h photoperiod fluorescent light [6].
7. It is necessary to check the level of PGM as the experiment progresses to maintain that the root system is always submerged in the solution.

3.2 RT-qPCR Quantification and Analysis

3.2.1 Collection of Samples

A typical fungal and/or plant gene expression analysis performed during the host plant colonization involves the quantification of gene expression in equivalent sections of plants inoculated with different fungal strains and/or cultured under different sets of conditions.

1. Select the time-points of interest for the collection of the samples. Take the plants out of the culture tubes, and wash the root system gently with distilled water.
2. Depending on the region of interest cut off the sections with a scalpel blade. Use a fresh scalpel blade for each condition to avoid cross-contamination.
3. Place the samples immediately in liquid nitrogen, and store at −80 °C until further use.

3.2.2 Isolation of RNA

Methods based in Tri-reagent have been widely used to isolate RNA from fungal and plant tissues due to the good yields obtained [15]. However, even better results may be obtained by using less time-consuming column-based methods, such as the one based on the Promega SV Total RNA Isolation System, with some modifications.

All the steps must be carried out keeping the samples on ice and using RNase-free consumables. All the solutions must be prepared with sterile RNase-free distilled water.

1. Remove samples from the −80 °C freezer. Grind 30 mg of each sample with mortar and pestle. Add liquid nitrogen and grind to obtain a fine dust.
2. Place the ground material in a 1.5 mL tube. Add 175 μL of Lysis Buffer to the sample. Mix thoroughly in a vortex for 10 s.
3. Add 350 μL of RNA Dilution Buffer. Pipette repeatedly to mix.
4. Heat the mixture at 70 °C for 3 min.
5. Centrifuge at 4 °C for 10 min at 14,000 × *g*. Transfer carefully the cleared lysate to a fresh pre-chilled 1.5 mL tube.
6. Add 200 μL of 95% ethanol and mix well.
7. Transfer the mixture to a pre-chilled Spin Basket Assembly. Centrifuge for 1 min as described in **step 5**. Discard the eluate.
8. Add 600 μL of RNA Wash Solution to the column, and centrifuge as described in **step 5**. Again, discard the eluate.
9. Prepare the DNase incubation mix (40 μL Yellow Core Buffer, 5 μL $MnCl_2$ 0.09 M, 5 μL DNase I), and mix gently (do not vortex!). The DNase incubation mix must be prepared right before its use. Apply the mixture onto the center of the membrane in the column, and incubate at room temperature for 15 min (*see* **Note 5**).
10. Add 200 μL of DNase Stop Solution to the column, and centrifuge as described in **step 5**.
11. Repeat **step 8**.
12. Add 250 μL of RNA Wash Solution to the column, and centrifuge for 2 min as described in **step 5**. Transfer the Spin Basket to a fresh elution tube.
13. Elute the RNA by adding 50 μL of nuclease-free water to the membrane. Centrifuge as described in **step 5**. Discard the column.
14. Store the RNA at −80 °C until use.

3.2.3 Determination of RNA Concentration and Quality

It is very important to assess the quality of the RNA samples, and accurately measure the RNA concentration. For both purposes absorbance readings may be obtained from very small volumes by using a Nanodrop spectrophotometer.

1. Set the Nanodrop to zero with 1 μL of distilled water. Configure the system for RNA measurements, and measure 1 μL of DEPC-treated water as the blank. Use 1–2 μL of each undiluted RNA sample. Clean the pedestal after measuring each sample. Measurement of blank samples between RNA samples is not needed.

2. Visualize your RNA sample by agarose gel electrophoresis in 1% gels following the MOPS Buffer/Formaldehyde protocol. Prepare a 1% agarose/formaldehyde gel by mixing 20 mL of 5× MOPS Buffer, 72 mL of DEPC-treated water, and 1 g of agarose.
3. Heat the mixture to boiling and cool to about 55 °C before the addition of 17.6 mL of 37% formaldehyde (use a fume hood!). Pour the gel and allow to cool. Use 1× MOPS Buffer for the electrophoresis.
4. Prepare the RNA sample by combining 1 μg of RNA and two volumes of RNA sample buffer and heating at 65 °C for 5 min. Immediately after heating, place the sample on ice for 2 min. Add RNA loading buffer before loading in the gel well (*see* **Note 6**).

3.2.4 Reverse Transcription from RNA to cDNA

There are several protocols to obtain cDNA from RNA. The one based on the PrimeScript™ RT Reagent (Takara Biosystems) has given excellent efficiency and yield when using RNA from samples collected from plant-fungal interactions [7, 16].

1. Mix 50 ng/μL of pure and DNase-treated RNA, 5× PrimeScript Buffer to a final concentration of 1×, 25 pmol of 50 μM Oligo dT primer, and 0.05 μg/μL of PrimeScript RT Enzyme Mix I in a RNase-free tube. Make up the final volume with sterile RNase-free water.
2. Incubate the mixture in a thermocycler under the following program: one step at 37 °C for 15 min, followed by one step at 85 °C for 5 s.
3. Store the cDNA at −80 °C until use (*see* **Note 7**).

3.2.5 Quantitative Real-Time PCR (RT-qPCR)

The protocol here makes use of SYBR ® Green. It presents the following advantages over other systems: user-friendly, low cost, and high efficiency in a wide range of samples.

The recommended protocol for the Step One Plus™ Real-Time PCR system has been adapted as detailed below. All the reactions are performed in a final volume of 10 μL. Perform the experiments keeping the reagents and tubes on ice. Avoid repeated freezing and thawing of samples and reagents to prevent degradation.

1. Mix 1× KAPA™ SYBR® FAST qPCR Kit Master Mix (2×) ABI Prism™, 600 nM forward primer, 600 nM reverse primer, and 1 μL of the transcription reaction obtained in Subheading 3.2.4. Bring to a total volume of 10 μL with sterile molecular grade water.

2. Transfer the reaction mixtures to the wells of a 96-well microtiter plate. Spin down the plate for 5–10 s, and keep it in the dark before placing in the thermocycler.
3. Run the qPCR reaction using the thermal profile recommended by the manufacturer when using SYBR® Green. For the Step One Plus™ Real-Time PCR System, this profile includes a first stage at 95 °C for 2 min, a second stage of 40 cycles of 95 °C for 3 s and 60 °C for 30 s, and a last stage to produce a melting or dissociation curve (0.3 °C increments every 5 s within a range of 55–95 °C) (*see* **Note 8**).

3.2.6 Analysis of Gene Expression

The analysis of the results is based on the Comparative Cycle Threshold method [17–19], which allows for a relative quantification of the target gene compared to the housekeeping gene. The Ct values obtained from the qPCR amplification curves are used to calculate the Relative Quantity of transcript (RQ), according to the following formula:

$$RQ = 2^{-Ct}$$

where $\Delta\Delta Ct = \Delta Ct^{\text{specific sample}} - \Delta Ct^{\text{reference sample}}$, being $\Delta Ct = Ct^{\text{target gene}} - Ct^{\text{housekeeping gene}}$ for any given sample.

The method requires statistically similar amplification efficiencies for target and housekeeping reference genes to ensure reliable quantification results. To this point, the PCR amplification efficiency of any pair of primers must be calculated as follows:

1. Prepare six serial dilutions, preferably 1/10 dilutions, of a genomic DNA sample with a known concentration to use as a template.
2. Prepare reaction mixtures for each dilution and primer pair designed to amplify a region of the genes whose expression has to be analyzed, according to the recommendations given above.
3. Run qPCR reactions as described in Subheading 3.2.5.
4. Analyze the efficiency of each pair of primers, and verify that it is greater than 96%.

In Fig. 2 we show the expression analysis of two fungal genes along the colonization of common bean plants inoculated with two different *F. oxysporum* f. sp. *phaseoli* strains (*see* **Note 9**).

3.3 Confocal Laser Microscopy

Inoculated plants are maintained in hydroponic cultures as described above before being used for microscopic analysis. At the time-points of interest, remove the plants from the tubes and wash the roots in SDW. Manually slice longitudinal and cross sections from root system, root crown, and hypocotyls. Use a razor blade to slice samples of approximately 1 mm in thickness. There are two

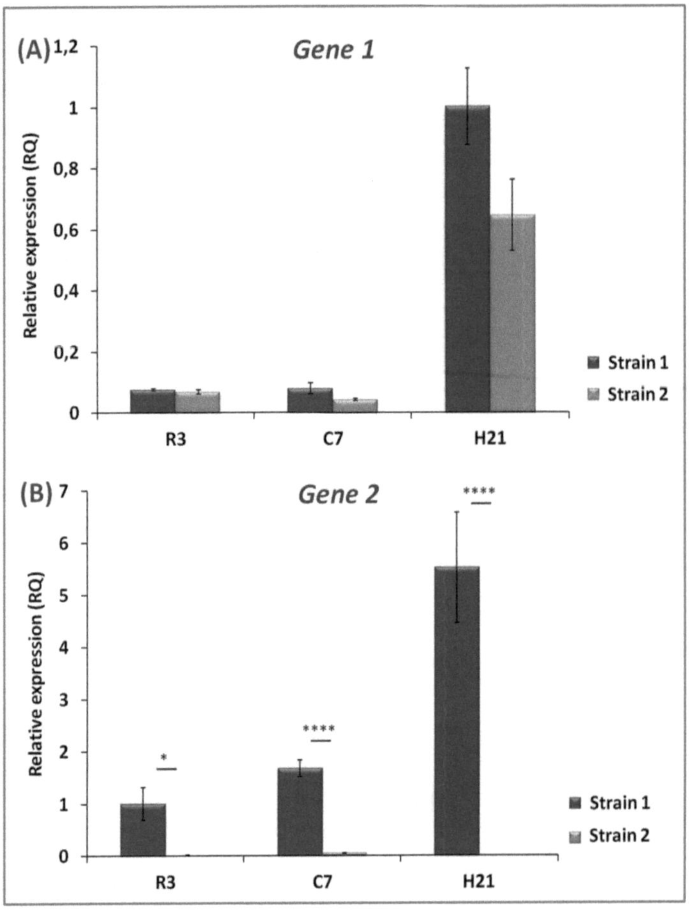

Fig. 2 RT-qPCR analysis of the expression of two fungal genes during common bean colonization with two *F. oxysporum* f. sp. *phaseoli* strains. The plant regions assayed and the time intervals after inoculation are indicated in the X axis (R3, root system at 3 dpi; C7, root crown at 7 dpi; H21, hypocotyls at 21 dpi). The expression ratios were normalized by using the fungal *EF1α* gene as housekeeping control. The value 1.0 was denoted for the transcript level of the reference samples: for *Gene* 1, Strain 1 in H21 (A); for *Gene 2*, Strain 1 in R3 (B). The levels of expression for each pair of measurements were tested using a *t*-test, and significant differences are indicated by * ($P < 0.05$) and **** ($P < 0.0001$)

staining alternatives depending on whether the plants were or were not infected with GFP-expressing fungal strains.

3.3.1 GFP-Expressing Fungal Strains

1. Place the plant sections on microscope slides, and add 20 μL of propidium iodide stock solution over the samples.
2. Incubate the plant sections for 15 min at RT, and remove the excess propidium iodide by washing the samples with SDW.

3.3.2 Non-GFP-Expressing Fungal Strains

The conjugate WGA Alexa Fluor™ 488 (Wheat Germ Agglutinin Alexa Fluor) provides bright green fluorescence when binding to the N-acetylglucosamine residues of the fungal cell wall. This property of the conjugate is used to visualize the fungal strains that do not express GFP. The protocol has been adapted from Poppe et al. [20].

1. Plant sections obtained as above are placed in sterile 10 mL tubes. Incubate for 16–24 h in 100% ethanol at 4 °C (ethanol should be changed 2–3 times during this incubation step).
2. Remove ethanol and add 5 mL of 10% KOH. Incubate at 85 °C for 5 min.
3. Wash the samples 4 times with 1× PBS. Transfer to 1.5 mL fresh tubes.
4. Add 1 mL of staining solution and then vacuum infiltrate 10 times for 2 min each time using a centrifugal vacuum concentration system (*see* **Note 10**).
5. Remove excess stain by washing with PBS. Samples can be visualized at this stage. However, best results are obtained after an incubation period of 12–24 h at 4 °C (*see* **Note 11**).

3.3.3 Confocal Laser Microscopy

1. Place the plant samples, stained either with propidium iodide or WGA Alexa Fluor, on fresh microscope slides. Submerge the samples in glycerol 30% (v/v) and gently squash with a cover glass.
2. Visualize immediately using a laser confocal microscope. GFP fluorescence is detected at 495–520 nm. WGA Alexa Fluor fluorescence is detected at 490–540 nm. Plant tissue fluorescence, either propidium iodide stained or produced by autofluorescence, is detected at 570–640 nm.

Figure 3 is an example of some confocal laser images from plants inoculated with *F. oxysporum* f. sp. *phaseoli*.

4 Notes

1. The addition of $CaSO_4 \cdot H_2O$ to the solution before autoclaving produces a precipitate. To avoid precipitation, prepare the $CaSO_4 \cdot H_2O$ solution, and sterilize by filtration. Add the filter sterilized $CaSO_4 \cdot H_2O$ solution to the autoclave sterilized PGM.
2. After autoclaving the solution will turn yellow, but this will not affect the quality of the buffer.
3. Avoid repeated freezing and thawing of the buffer. The buffer can be stored for up to 6 months at −20 °C.

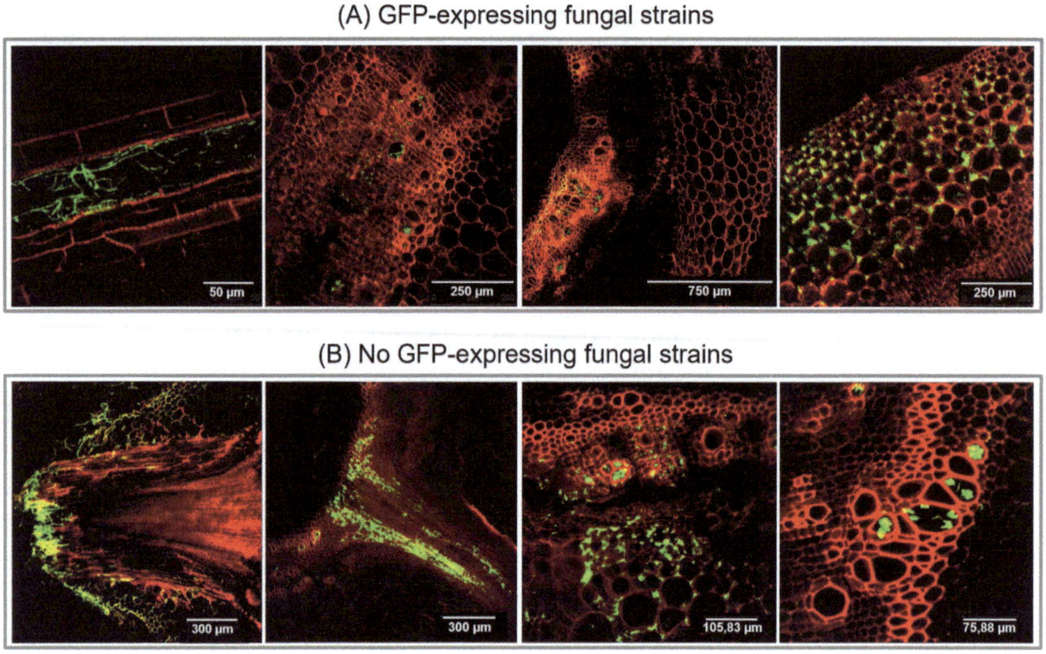

Fig. 3 Common bean colonization by *F. oxysporum* f. sp. *phaseoli* strains. Confocal laser microscopy was used to obtain images of different plant regions at different time-points. (**a**) Sections from plants inoculated with GFP-expressing fungal stained with propidium iodide. (**b**) Sections from plants inoculated with no GFP-expression fungal strains stained with WGA-propidium iodide

4. It is highly recommended to use a high-grade glycerol to avoid ribonuclease activity.
5. Although this is an optional step, it is recommended to perform an additional treatment with DNase to completely remove traces of contaminating genomic DNA from the sample.
6. In Nanodrop measurements, a ratio of the absorbances at 260/280 nm around 2 is accepted as pure RNA. Lower values indicate the presence of proteins or other contaminants. The ratio at 260/230 nm has to be higher than the 260/280 ratio for pure RNA. Values over 2.2 or less than 1.8 indicate the presence of contaminants. In agarose gels, intact RNA is observed as two bands around 2 kb and 4 kb corresponding to the 18 S and 28 S ribosomal RNA, respectively. Degraded RNA appears as a lower molecular weight smear.
7. It is highly recommended to check the absence of contamination with genomic DNA in the cDNA reaction by running a Polymerase Chain Reaction (PCR) aimed to the amplification of an intron-containing region selected from a housekeeping gene.

8. The melting curve is useful to verify the presence of only one specific peak corresponding to the amplification of a single amplicon. It is highly recommended to verify this point, and also the correct size of the amplicon, by agarose gel electrophoresis in 2.5% (w/V) gels.
9. In order to obtain reliable results, it is recommended to analyze three sets of samples from different inoculation assays (biological replicates), two independent cDNA preparations per biological replicate, and three replicas of each cDNA. Use the data from all the qPCR reactions to calculate means and standard deviations.
10. The most critical step in this procedure is the number and duration of vacuum infiltration pulses with the Staining Solution. This step should be optimized for each plant-fungal interaction.
11. The duration of the incubation step at 4 °C prior to microscope visualization may also be varied for each plant-fungal interaction.

Acknowledgments

We are indebted to Michael R. Thon for critical reading of the manuscript and many suggestions. This work was supported by Ministerio de Ciencia e Innovación grant PID2019-110605RB-100 (Spain).

References

1. Sacristán S, García-Arenal F (2008) The evolution of virulence and pathogenicity in plant pathogen populations. Mol Plant Pathol 9:369–384
2. de Vega-Bartol JJ, Martín-Dominguez R, Ramos B et al (2011) New virulence groups in *Fusarium oxysporum* f. sp. *phaseoli*: the expression of the gene coding for the transcription factor ftf1 correlates with virulence. Phytopathology 101:470–479
3. Bishop CD, Cooper RM (1983) An ultrastructural study of root invasion in three vascular wilt diseases. Physiol Plant Pathol 22:15–27
4. Heath MC (2000) Advances in imaging the cell biology of plant-microbe interactions. Annu Rev Phytopathol 38:443–459
5. Michielse CB, Rep M (2009) Pathogen profile update: *Fusarium oxysporum*. Mol Plant Pathol 10:311–324
6. Jiménez-Fernández D, Landa BB, Kang S et al (2013) Quantitative and microscopic assessment of compatible and incompatible interactions between chickpea cultivars and *Fusarium oxysporum* f. sp. *ciceris* races. PLoS One 8:e61360
7. Niño-Sánchez J, Tello V, Casado-del Castillo V et al (2015) Gene expression patterns and dynamics of the colonization of common bean (*Phaseolus vulgaris* L.) by highly virulent and weakly virulent strains of *Fusarium oxysporum*. Front Microbiol 6:605–614
8. García-Sánchez A, Martín-Rodrigues N, Ramos B et al (2010) fost12, the *Fusarium oxysporum* homolog of the transcription factor Ste12, is upregulated during plant infection and required for virulence. Fungal Genet Biol 47:216–225
9. Huggett J, Dheda K, Bustin S et al (2005) Real-time RT-PCR normalisation; strategies and considerations. Genes Immun 6:279–284
10. Hadas Y, Goldberg I, Pines O et al (2007) Involvement of gluconic acid and glucose

oxidase in the pathogenicity of *Penicillium expansum* in apples. Phytopathology 97:384–390

11. Yedidia I, Benhamou N, Chet I (1999) Induction of defense responses in cucumber plants (*Cucumis sativus* L.) by the biocontrol agent *Trichoderma harzianum*. Appl Environ Microbiol 65:1061–1070
12. Pastor-Corrales MA, Abawi GS (1987) Reactions of selected bean germ plasms to infection by *Fusarium oxysporum* f. sp. *phaseoli*. Plant Dis 71:990
13. Alves-Santos FM, Cordeiro-Rodrigues L, Sayagués JM et al (2002) Pathogenicity and race characterization of *Fusarium oxysporum* f. sp. *phaseoli* isolates from Spain and Greece. Plant Pathol 51:605–611
14. Abawi GS (1990) Root rots of beans in Latin America and Africa: diagnosis, research methodologies, and management strategies. CIAT, Cali
15. De Vega-Bartol JJ, Tello V, Niño-Sánchez J et al (2013) Quantitative PCR analysis of double-stranded RNA-mediated gene silencing in fungi. In: Gupta VK, Tuohy MG, Ayyachamy M, Turner KM, O'Donovan A (eds) Laboratory protocols in fungal biology: current methods in fungal biology. Springer New York, New York, pp 279–287
16. Niño-Sánchez J, Casado-Del Castillo V, Tello V et al (2016) The FTF gene family regulates virulence and expression of SIX effectors in *Fusarium oxysporum*. Mol Plant Pathol 17:1124–1139
17. Bustin SA (2000) Absolute quantification of mRNA using real-time reverse transcription polymerase chain reaction assays. J Mol Endocrinol 25:169–193
18. Pfaffl MW (2001) A new mathematical model for relative quantification in real-time RT-PCR. Nucleic Acids Res 29:e45
19. Livak KJ, Schmittgen TD (2001) Analysis of relative gene expression data using real-time quantitative PCR and the 2(-Delta Delta C (T)) method. Methods 25:402–408
20. Poppe S, Dorsheimer L, Happel P et al (2015) Rapidly evolving genes are key players in host specialization and virulence of the fungal wheat pathogen *Zymoseptoria tritici* (*Mycosphaerella graminicola*). PLoS Pathog 11:e1005055

Chapter 13

Time-Lapse Imaging of Root Pathogenesis and Fungal Proliferation Without Physically Disrupting Roots

Hye-Seon Kim, Sook-Young Park, Seogchan Kang, and Kirk J. Czymmek

Abstract

Microscopic observation of root disease onset and progression is typically performed by harvesting different plants at multiple time points. This approach prevents the monitoring of individual encounter sites over time, often mechanically damages roots, and exposes roots to unnatural conditions during observation. Here, we describe a method developed to avoid these problems and its application to study *Fusarium oxysporum-Arabidopsis thaliana* interactions. This method enabled three-dimensional, time-lapse imaging of both *A. thaliana* and *F. oxysporum* as they interact via the use of confocal and multi-photon microscopy and facilitated inquiries about the genetic mechanism underpinning Fusarium wilt.

Key words *Arabidopsis thaliana*, Calcium indicator, Confocal microscopy, Fluorescent protein, *Fusarium oxysporum*, Root colonization, Soilborne disease

1 Introduction

Soilborne diseases threaten global food security and sustainability, but available control strategies are limited. Pre-plant soil fumigation has been effective for controlling soilborne diseases, especially for high-value crops. However, since the removal of methyl bromide (MB) due to its effect on the Earth's ozone layer [1], some soilborne diseases, including *Fusarium* wilt, have reemerged. However, MB alternatives are not as effective, calling for novel strategies to help mitigate their impact without continuously relying on fumigants. An improved understanding of the nature and mechanism of plant-pathogen interactions in the rhizosphere is crucial to guide the development and deployment of such strategies.

Unlike foliar and fruit pathogens that cause conspicuous symptoms that can be monitored over time, root pathogens pose several technical challenges to following them via microscopy. Root disease onset and progression has been typically documented by analyzing different plants harvested at multiple time points during the disease

Jeffrey Coleman (ed.), *Fusarium wilt: Methods and Protocols*,
Methods in Molecular Biology, vol. 2391, https://doi.org/10.1007/978-1-0716-1795-3_13,

cycle. This approach eliminates the possibility of following individual encounter sites over time. Time-lapse imaging of the same sites has revealed many important aspects of foliar pathogenesis [2]. For roots, extreme care must be exercised to avoid mechanical damage during harvesting because wounding potentially creates sites for pathogen entry and affects root physiology. Even carefully collected roots experience unnatural conditions due to disruption of the local environment during extraction, transfer for imaging, and microscopic observation, likely affecting plant-pathogen interactions.

Here, we introduce a method designed for time-lapse imaging of root pathogenesis by *Fusarium oxysporum* while avoiding these problems. This method was applied to investigate the mechanism of *Arabidopsis thaliana-F. oxysporum* interactions [3, 4]. *Arabidopsis thaliana* has been extensively utilized for studying plant biology, including defense mechanisms against diverse pathogens including *F. oxysporum* [5, 6], and the nature and mode-of-action of pathogen factors involved in plant infection [4, 7]. Its small foliage and root size, rapid growth, and the vast collection of mutants and other resources are highly desirable attributes for investigating the molecular and cellular mechanism underpinning root pathogenesis. Two main components of our method include seedlings of *A. thaliana* grown in commercially available coverslip-bottomed chambers (Fig. 1) and *F. oxysporum* strains expressing fluorescent proteins (FPs). Tracking a specific pathogen that colonizes roots is complicated because roots are already associated with vastly diverse microorganisms [8, 9]. Genetic transformation of *F. oxysporum* to express FPs with different spectral properties has facilitated in vivo imaging of root pathogenesis with unequivocal determination of the fungus and its locality within different root tissues and structures without requiring addition of exogenous stains [3, 4].

Via the method described below, three-dimensional (3-D) images from individual encounter sites between *A. thaliana* and *F. oxysporum* could be acquired repeatedly over several days without physical manipulation or retrieval of infected plants [3]. Resulting data revealed fungal structures associated with infection, location of fungal colonization and penetration, the exact timing and voracity of infection, and host cellular responses (e.g., autofluorescence as a potential indicator for chemical defense, cell collapse presumably caused by phytotoxic molecules secreted by *F. oxysporum*) to infection. Besides, the rate of fungal ingress within the vascular tissue and changes in surrounding root cortex cells in response to molecules secreted by advancing fungal hyphae could be documented, providing invaluable insights into how *F. oxysporum* affects host cells inside the vascular tissue. We also applied this method to image spatial and temporal dynamics of subcellular Ca^{2+} (i.e., "Ca^{2+} signature") in *F. oxysporum* as its hyphal tip interacted with the root surface and advanced within vascular tissues of *A. thaliana* [10]. This imaging method can provide an unprecedented vantage

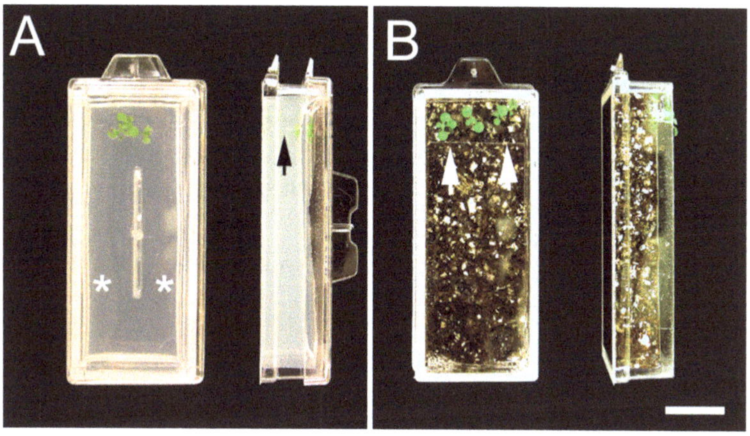

Fig. 1 Plant growth chamber with coverglass bottom. (**a**) Top and side view of 12-day-old *Arabidopsis* seedlings in Nalgene Nunc No. 1.5 chamber with MS agar. Asterisks denote location of fungal inoculation, and arrow shows position of root extending toward the coverglass through the agar. (**b**) Top and side view of 12-day-old *Arabidopsis* seedlings in the same system with a 50:50 mixture of potting soil and vermiculite amended with a modified MS medium. Arrows denote top edge of a plastic coverslip used to hold soil in position. Scale bar = 10 mm. (Reproduced with permission from Czymmek et al. [3])

of the root disease process while minimizing artifacts caused by classical methodologies and should provide a critical frame of reference for researchers working with other plant-soil microbe interactions.

2 Materials

2.1 Biological Materials

1. Seeds of *A. thaliana* ecotype Columbia (Col-0) and Cape Verde Island (Cvi-0).
2. Fungal strains: *Fusarium oxysporum* O-685, a strain isolated from diseased cabbage, was acquired from the Penn State *Fusarium* Research Center. Spores of O-685 and its transformants in 20% glycerol are kept in a −80 °C freezer for long-term storage and are revitalized before experiments.
3. *Agrobacterium tumefaciens* strain EHA105.
4. Plasmid 615 contains the ZsGreen gene, which encodes a green FP, driven by the *F. verticilliodes* translation elongation factor 1α (*EF-1α*) gene promoter along with the hygromycin B resistance gene driven by the *Cochliobolus heterostrophus* glyceraldehyde 3-phosphate dehydrogenase gene promoter [11].
5. Plasmid pSK2789 [10] carries the coding region of YC3.60, which encodes a genetically encoded Ca^{2+} indicator Cameleon, under the *EF-1α* gene promoter in pBHt2, a binary vector for

fungal transformation using *A. tumefaciens*. pBHt2 contains a gene conferring resistance against hygromycin B as a selection marker in the T-DNA region.

6. pGreenII containing the coding region of YC3.60 under the control of the cauliflower mosaic virus 35S promoter [12]. This binary vector carries a kanamycin resistance gene as a selection marker.
7. Lysing enzymes from *Trichoderma harzianum* (Sigma-Aldrich L1412).

2.2 Culture Media and Solutions

All media noted below were autoclaved for 15 min at 121 °C on liquid cycle.

1. Murashige and Skoog (MS) agar [13]: Prepare by adding 4.4 g MS salt and vitamin mixture (Sigma-Aldrich M0404), 0.5 g MES, and 0.8% Phytagar to 1 L deionized H_2O (adjust pH to 5.7 using 1 M KOH).
2. Arabidopsis Nutrient (AN) agar: Contains 5 mM KNO_3, 2.5 mM KH_2PO_4, 2 mM $MgSO_4$, 2 mM $Ca(NO_3)_2$, 0.005% sequestrine, 70 μM H_3BO_3, 14 μM $MnCl_2$, 0.5 μM $CuSO_4$, 1 μM $ZnSO_4$, 0.2 μM Na_2MoO_4, 10 μM NaCl, 10 μM $CoCl_2$, and 0.7% agar.
3. A mixture of vermiculite with either potting soil or washed playground sand (50:50) is used for growing *A. thaliana* seedlings employed for soil-based microscopy imaging of root pathogenesis. The mixture is sterilized via autoclave and supplemented with MS medium (without Phytagar) (*see* **Note 1**).
4. Carboxymethyl cellulose (CMC) broth: Dissolve 15 g of CMC, 1 g of yeast extract, 0.5 g of $MgSO_4$, 1 g of NH_4NO_3, and 1 g of KH_2PO_4 in 1 L of water. This medium was used for producing *F. oxysporum* microconidia in large scale.
5. Complete medium (CM): Prepare by adding 1 g KH_2PO_4, 0.5 g $MgSO_4 \cdot 7\ H_2O$, 0.5 g KCl, 20 g agar, 0.2 mL trace element solution, 30 g sucrose, 2 g $NaNO_3$, 2.5 g N-Z Amine, 1 g yeast extract, 10 mL vitamin stock solution in 1 L ddH_2O.
6. Trace element solution: Add 5 g citric acid, 5 g $ZnSO_4 \cdot 6\ H_2O$, 1 g $Fe(NH_4)_2(SO_4)_2 \cdot 6\ H_2O$, 250 mg $CuSO_4 \cdot 5\ H_2O$, 50 mg $MnSO_4$, 50 mg H_3BO_3 (boric acid), 50 mg $Na_2MoO_4 \cdot 2\ H_2O$ in ~90 mL distilled water. Completely dissolve citric acid before adding other ingredients to prevent salts from precipitating. Bring the volume to 100 mL before sterilizing via filtration. Store the solution in 4 °C.
7. Vitamin stock solution: 1 L solution in ethanol:H_2O (50:50) contains 4 g inositol, 200 mg calcium pantothenate, 200 mg choline chloride, 100 mg thiamine, 75 mg pyridoxine, 75 mg

nicotinamide, 50 mg ascorbic acid, 30 mg riboflavin, 5 mg *p*-aminobenzoic acid, 5 mg folic acid, 5 mg biotin. The solution was filter-sterilized and stored in 4 °C (dark bottle or bottle wrapped in aluminum foil). Store in the dark because some of the vitamins decompose when exposed to light.

8. 0.5× Potato Dextrose Agar (PDA): 19.5 g PDA in 1 L ddH_2O. Pour into Petri plates and allow to solidify.
9. 0.75× PDA: 29.25 g PDA in 1 L ddH_2O.
10. 70% and 95% ethanol.
11. 6% sodium hypochlorite solution (CLOROX).
12. Water agar (0.1% and 1%).
13. Carnation leaves sterilized through gamma irradiation (acquired from the Penn State *Fusarium* Research Center).
14. PTC (40% PEG 1000 in 20% sucrose, 50 mM Tris, pH 8.0, 50 mM $CaCl_2$).
15. Triton X-100.
16. Pro-Mix (Premier Tech Horticulture).
17. Silwet L-77, an organosilicone surfactant (Plantmedia).
18. Hoagland solution: 5 mL 1 M KNO_3, 2.5 mL 1 M KH_2PO_4, 2 mL 1 M $MgSO_4$, 2 mL 1 M $Ca(NO_3)_2$, 1 mL micronutrient mix (2.86 g H_3BO_3, 1.81 g $MnCl_2 \cdot 4\ H_2O$, 0.22 g $ZnSO_4 \cdot 7\ H_2O$, 0.08 g $CuSO_4 \cdot 5\ H_2O$, and 0.02 g $H_2MoO_4 \cdot\ H_2O$ in 1 L ddH_2O), and 1 mL Fe-EDTA solution (10.4 g EDTA-Na_2, 7.8 g $FeSO_4 \cdot 7\ H_2O$ and 56.1 g KOH in 1 L ddH_2O) in 1 L ddH_2O (*see* **Note 2**).

2.3 Equipment and Supplies

1. Single-welled Lab-Tek® Nalgene Nunc chambered coverglass system.
2. 22 × 60 mm Thermanox plastic coverslip.
3. Scissors.
4. 100 × 15 mm petri dishes for culturing *F. oxysporum*.
5. Flasks for producing *F. oxysporum* spores using CMC.
6. Aracon (Arasystem), consisting of Aratube (plastic sheet that can be folded into a tube) and Arabase (inverted plastic cone that guides the growing plant in an Aratube and collects the falling seed), was used to harvest seeds from individual *A. thaliana* transformants without cross pollination and seed contamination.
7. Environmental Control system.

2.4 Microscope Hardware and Software

1. GFP 470 filter on a Zeiss M^2Bio Microscope.
2. Zeiss LSM 510 NLO confocal scanhead on an inverted Axio-Vert 200 M microscope with LSM 510 AIM software (Rel.

3.2). The system should be equipped with a Coherent Enterprise MIRA Verdi 5 W Ti:Sapphire femtosecond tunable laser, 25 mW argon ion laser with 458 nm, 488 nm, and 514 nm lines, 1 mW 543 HeNe laser, and 5 mW 633 HeNe laser (*see* **Note 3**).

3. LSM 510 emission filters: 505 longpass, 500–550 bandpass emission filter (for ZsGreen), 560 longpass emission filter (for autofluorescence of soil and plant cell walls), 400–460 bandpass (for autofluorescence of plant cell walls, vacuolar contents, and xylem elements), and 465–510 nm bandpass (CFP) and 530 nm longpass (YFP) emission filters (for Förster resonance energy transfer (FRET)-based imaging of subcellular Ca^{2+}).
4. Transmitted light and epifluorescence modes with filter cubes for GFP, RFP, and Cy5 on the LSM510 NLO confocal microscope.
5. Zeiss 5× Fluar (0.25 NA), 10× Plan-Apochromat (0.45 NA), 20× Plan-Apochromat (0.75 NA), 40× oil Plan-NeoFluar (1.3 NA), and 40× C-Apochromat (1.2 NA) water immersion objective lenses.
6. MultiTime Macro (ver. Q) in conjunction with a motorized x-y stage, a data acquisition software that allows multi-location, z-stack, time-series, and autofocus routines (*see* **Note 3**).

3 Methods

3.1 Transformation of F. oxysporum

Strain O-685 is transformed using plasmid 615, which contains the ZsGreen gene, and pSK2789, a binary vector containing Cameleon YC3.60, a genetically encoded Ca^{2+} indicator, via protoplasts [3] and *Agrobacterium tumefaciens* [14], respectively. *A. tumefaciens* strain EHA105 carrying plasmid pSK2789 was used to transform O-685 [10]. Because we already published a very detailed protocol for *F. oxysporum* transformation using *A. tumefaciens* [15, 16], we only describe the generation of O-685 transformants expressing ZsGreen here.

1. Use a sterile 200 μL pipette tip to inoculate ~20 μL glycerol stock of *F. oxysporum* spores at the center of a 0.5× PDA plate, and incubate at 25 °C for 6 days.
2. Harvest the conidia from the culture and transfer to 200 mL CM.
3. Incubate the culture on a shaker at 25 °C and 100 rpm for 5 days.
4. Collect the mycelia by filtration through a Whatman filter paper #4, wash with sterile water, and weigh to estimate the amount of lysing enzymes needed to digest the cell wall.

5. Cell wall digestion was performed as previously described [17] with some modifications. The mixture of mycelia and the enzyme (50 mg/mL lysing enzymes) in 0.7 M NaCl was incubated at 30 °C with gentle shaking. Monitor the progression of protoplast formation using a microscope.
6. After filtering the solution via two layers of Miracloth, harvest the protoplasts via centrifugation. After two washes with STC buffer (1.2 M sorbitol, 10 mM Tris–HCl pH 7.5, 10 mM $CaCl_2$), the protoplasts were diluted to 1×10^8 protoplasts per ml in STC.
7. Gently mix 20 μL of plasmid 615 [11] with 140 μL of protoplasts, and incubate at RT for 15 min.
8. Add 1 mL of PTC into the vector DNA/protoplast solution, and then incubate for 20 min at RT.
9. Mix the protoplasts with 4 mL 0.7% molten low melting temperature (LMP) agarose (50 °C) in CM with 20% sucrose. Overlay the mixture onto CM plates supplemented with 20% sucrose, and incubate overnight at 28 °C.
10. Overlay the plates with 10 mL CM with 0.7% LMP agarose and 100 μg/mL hygromycin B to each plate, and incubate at 28 °C until hygromycin B-resistant colonies emerge.
11. Transfer these colonies to individual CM plates containing 100 μg/mL hygromycin B for sporulation.
12. Spread conidia from individual putative transformants on CM containing 100 μg/mL hygromycin B, and incubate them overnight for germination.
13. Isolate germinated conidia using a stereomicroscope, transfer them to new medium for spore production, and preserve them in 20% glycerol solution at −80 °C until their fluorescence and phenotypes are checked and confirmed.
14. Screen *F. oxysporum* transformants using a GFP 470 filter on a Zeiss M^2Bio microscope, and score based on relative intensity.
15. For a few transformants displaying strong fluorescence, evaluate their virulence using *A. thaliana* ecotype Cvi-0, which is highly susceptible to O-685, and growth characteristics with O-685 as a control to confirm the transformation process did not affect these traits. Analyze the transformants via genomic Southern to determine the copy number of the ZsGreen gene.

3.2 Transformation of *A. thaliana*

The YC3.60 construct under the control of the 35S promoter in pGreenII [12] is used to transform *A. thaliana* ecotype Cvi-0. A floral dip procedure, a simple method that has been widely used to transform *A. thaliana* using *A. tumefaciens* [18], is provided.

1. Grow *A. thaliana* in the presence of Aracon (*see* Subheading 2.2).
2. Dip the developing flowers into a solution containing *A. tumefaciens* EHA105 carrying pGreenII with YC3.60, 5% sucrose, and 0.5 mL/L of Silwet L-77 [18].
3. After dipping, the plants are placed in a plastic tray and covered with a tall clear plastic dome to maintain humidity.
4. After placing the tray in a dark chamber for 24 h, return the transformed plants to a growth chamber set at 24 °C under a 16-h light/8-h dark photoperiod.
5. After growing the plants for 3–5 weeks until siliques were brown and dry, harvest seeds collected at the bottom of Arabase.
6. Select transformants by plating the sterilized seeds on MS medium with vitamins, 0.8% Phytagel, 30 μg/mL kanamycin, and 1% (w/v) sucrose.
7. Incubate the plates at 21 °C under continuous light conditions for 7 days.
8. Transfer the seedlings to plastic pots filled with potting mix for seed production.
9. Check several independent lines for fluorescence (yellow and cyan). Seeds are harvested from a mother plant that had been transformed with pGreenII with YC3.60. Check level of fluorescence of T1 and T2 transformant seedlings grown on MS media under selection with kanamycin under the fluorescence microscope, especially the *Arabidopsis* root.
10. The transformants with an adequate level of fluorescence are self-fertilized for three generations to obtain homozygous lines.

3.3 Setup, Inoculation, and Maintenance of Plants in Optical Growth Chambers

We infected *A. thaliana* seedlings grown in single-welled Lab-Tek® Nalgene Nunc chambered coverglass system with O-685 transformed with the ZsGreen gene or Cameleon (Fig. 1). Two different substrates were used to grow *A. thaliana* before fungal inoculation [3].

1. To activate O-685 transformants, use a sterile 200 μL pipette tip to inoculate frozen spores in 20% glycerol at the center of 1% water agar supplemented with irradiated carnation leaf pieces, and incubate at 25 °C for 6 days.
2. Rinse *F. oxysporum* spores from the plates with sterile water, filter the spore suspension through sterile cheesecloth followed by Miracloth, harvest the spores by centrifugation, wash them with sterile water, and adjust the concentration to 10^4 conidia/mL using a hemacytometer (*see* **Note 4**).

3. For agar-based imaging of *A. thaliana* roots: Add 4 mL of melted MS agar to each coverglass-bottomed chamber, and allow to solidify (Fig. 1a).
4. For soil-based imaging: Add ~4 mL of a 50:50 (v/v) mixture of vermiculite and potting soil or washed sand supplemented MS media (Fig. 1b). Cut a sterile 22 × 60 mm Thermanox plastic coverslip to ~22 × 50 mm using a sterilized scissor, and place it on the lower portion of the chamber (arrows in Fig. 1b) to hold the soil in position during re-orientation of the chamber (see below).
5. Sterilize *A. thaliana* seeds by washing with a bleach:dH_2O (50:50) solution with 0.01% Triton X-100 for 7 min followed by four washes with dH_2O for 1 min each. Keep the sterilized seeds in 0.1% water agar.
6. Place four sterilized seeds in a row approximately 2 mm from one end of the rectangular chamber (Fig. 1).
7. After wrapping the chamber with Parafilm to maintain humidity, keep it at 4 °C for 72 h.
8. Move the chamber to an environmental control system set at a constant temperature of 21 °C, 13 h dark and 11 h light cycle with a light intensity of 65 μmol/m^2/s.
9. When placing the chamber, maintain it in a horizontal orientation until all primary roots have reached the coverslip.
10. When all roots have grown to the coverslip, inoculate the chamber with fungal spores ($\leq$10 conidia/μL; 5 μL per inoculation point) by pushing the pipette tip through the agar or soil until you feel the tip come in contact with the coverslip. Do this on both sides of the chamber ~2/3 of the way along the chamber from the seedlings (see asterisks in Fig. 1a for position).
11. Place the inoculated chamber back into the incubator, but reorient it in a near vertical position (lean against the incubator interior wall), which allows subsequent root growth to occur adjacent to the coverslip surface for best imaging.
12. When hyphae are in close proximity to the root (~100 μm or less), 3D confocal imaging can proceed as described below (*see* **Note 5**).

3.4 Confocal Microscopy of Root Pathogenesis

Time-lapse documentation of root pathogenesis using confocal microscopy provides many novel insights of plant-fungal pathogenesis, as it is possible to track disease progression of individual interaction sites in an unperturbed optical chamber. The following includes the strategy needed with good working guidelines and resulting representative datasets of interaction sites in soil (Figs. 2 and 3) or agar medium (Figs. 4 and 5). Additionally, leveraging the

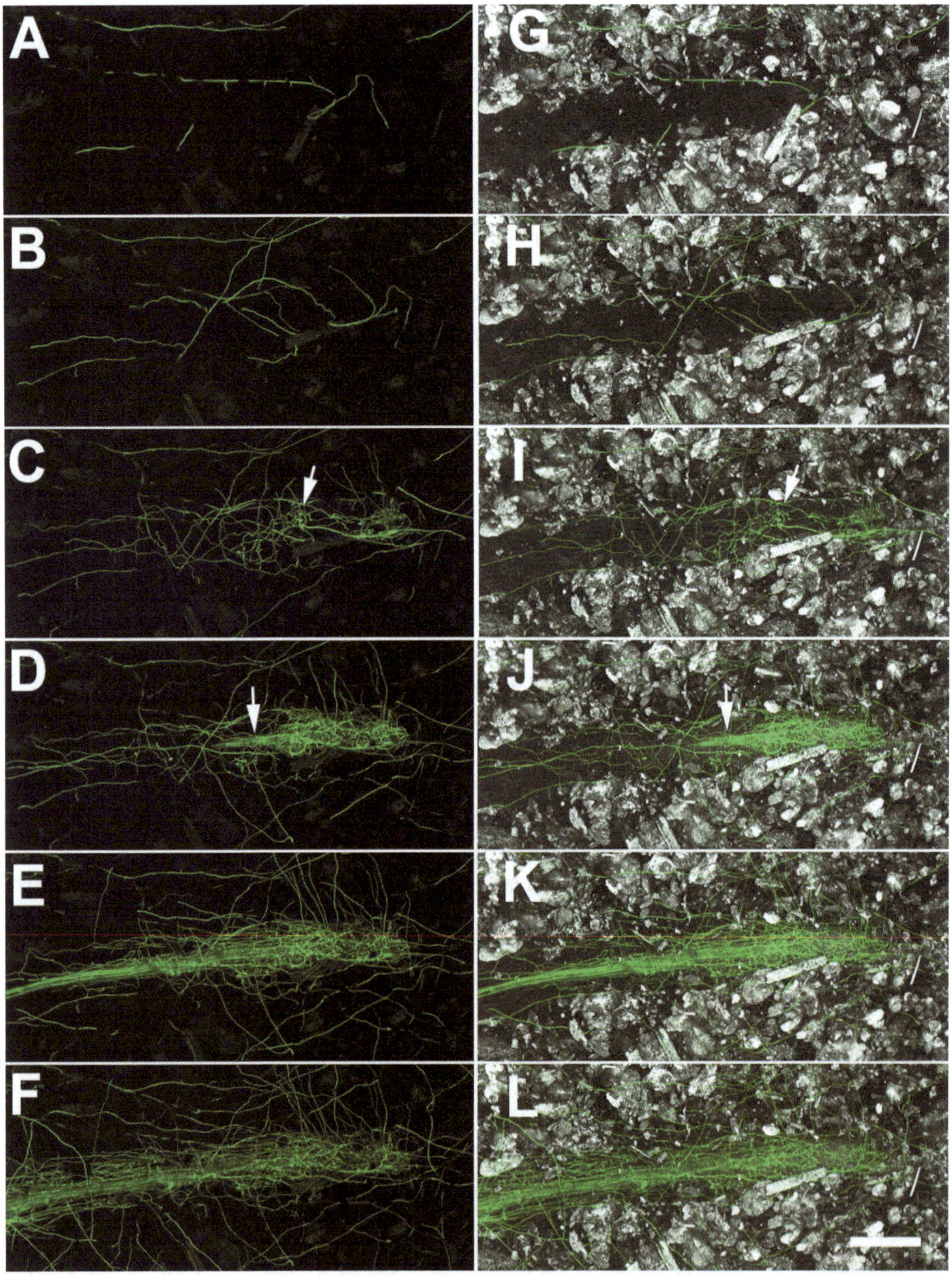

Fig. 2 4-D time-lapse images of *A. thaliana* ecotype Cvi-0 in soil 4–6 days post inoculation with *F. oxysporum*. Early colonization of the primary root tip by *F. oxysporum* strain 0-685 expressing ZsGreen (green) was primarily along anticlinal walls parallel with the long access of the root (**a**,**g**:**b**,**h**). Early infection was also characterized by localized hot spots representing enlarged fungal hyphae expressing ZsGreen (arrow, **c**,**i**). Several hours later the meristematic and vascular tissues were breached (arrow, **d**,**j**). Vascular invasion continued rapidly (**e**,**k**) followed by adjacent tissues (**f**,**l**). (**a**–**f**) Hyphae only (green). (**g**–**l**) hyphae (green) plus soil/autofluorescence (grayscale). Frames (**a**,**g**) $t = 0$ h, (**b**,**h**) $t = 6.1$ h, (**c**,**i**) $t = 23.18$ h, (**d**,**j**) $t = 29.28$ h, (**e**, **k**) $t = 34.16$ h, (**f**,**l**) $t = 43.92$ h. Scale bar is 100 μm. (Reproduced with permission from Czymmek et al. [3])

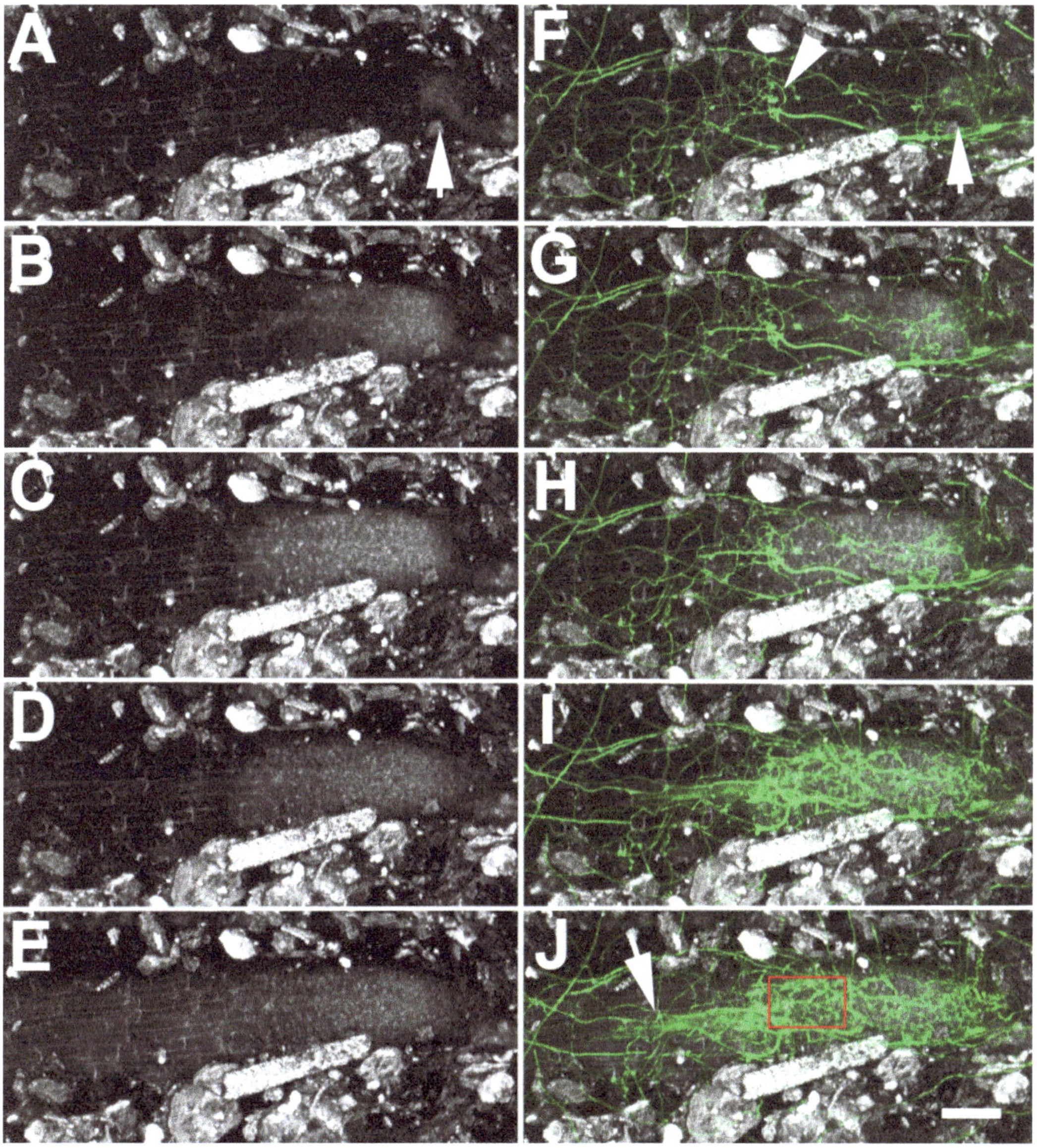

Fig. 3 4-D time-lapse of *A. thaliana* ecotype Cvi-0 in soil 4–5 days post inoculation with *F. oxysporum*. These images were collected from the primary toot tip shown in Fig. 2. One day following initial fungal contact, a wave of plant autofluorescence originated at the apex of the root tip (**a**,**f** arrow). Bright swollen hyphae were documented distal from the root apex (**f**, arrowhead) but apparently not the direct cause of this wave of plant autofluorescence. Within 1 h of initiation of the autofluorescence wave, the entire meristem exhibited a similar response (**a**–**c**, **f**–**h**). Approximately 4 h later, a second wave of plant autofluorescence was observed (cf. **d**,**i**, and **e**,**j**) extending into the elongation zone as the vascular tissues (arrow) were presumably breached. Frame (**a**,**f**) t = 22.57 h, (**b**,**g**) *t* = 23.18 h, (**c**,**h**) *t* = 23.79 h, (**d**,**i**) *t* = 28.06 h, (**e**,**j**) *t* = 28.67 h. Seventy-two frames acquired at 36.67-min interval. Time-lapse duration 43.38 h. Scale bar = 50 μm. (Reproduced with permission from Czymmek et al. [3])

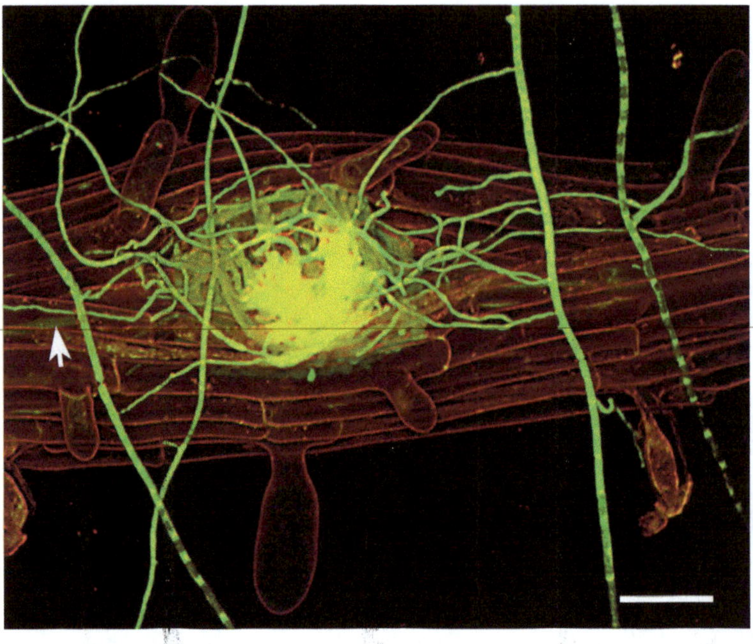

Fig. 4 High-resolution 3-D maximum intensity projection of one time point of *A. thaliana* infected by *F. oxysporum*. The image was captured from *A. thaliana* ecotype Cvi-0 (grown using agar) 5 days post inoculation with *F. oxysporum* strain 0-685 expressing ZsGreen. Its hyphae (green) targeted, encapsulated, and penetrated emerging lateral roots, subsequently gaining access to xylem vessels (arrow). Plant cell wall autofluorescence (red). Scale bar = 20 μm

FP-tagged *F. oxysporum* strain along with the autofluorescence of plant and soil structures allows for a more definitive and comprehensive qualitative and quantitative understanding of host-pathogen interactions and host response over time [3].

1. Observe the Lab-Tek chambers through microscope oculars for plant-fungal interaction sites with 5× or 10× objective lens in fluorescence imaging mode and a GFP filter set.
2. Identify areas of imminent or early fungal encroachment or penetration, and then switch to a 20× (Figs. 2 and 3) or 40× (Figs. 4 and 5) objective lens depending on the required field-of-view or cellular detail (*see* **Note 6**).
3. Depending on the type of data desired, images should be acquired using the appropriate laser and filter combinations as follows: 488 nm laser line (for excitation of ZsGreen) with 505 longpass or 500–550 bandpass emission filter (for ZsGreen), the 543 nm line of a helium:neon laser with a 560 longpass emission filter (for autofluorescence of soil and plant cell walls) (Figs. 2–4), or the 750 nm line of a 5 W pumped MIRA-900 titanium:sapphire laser with a 400–460

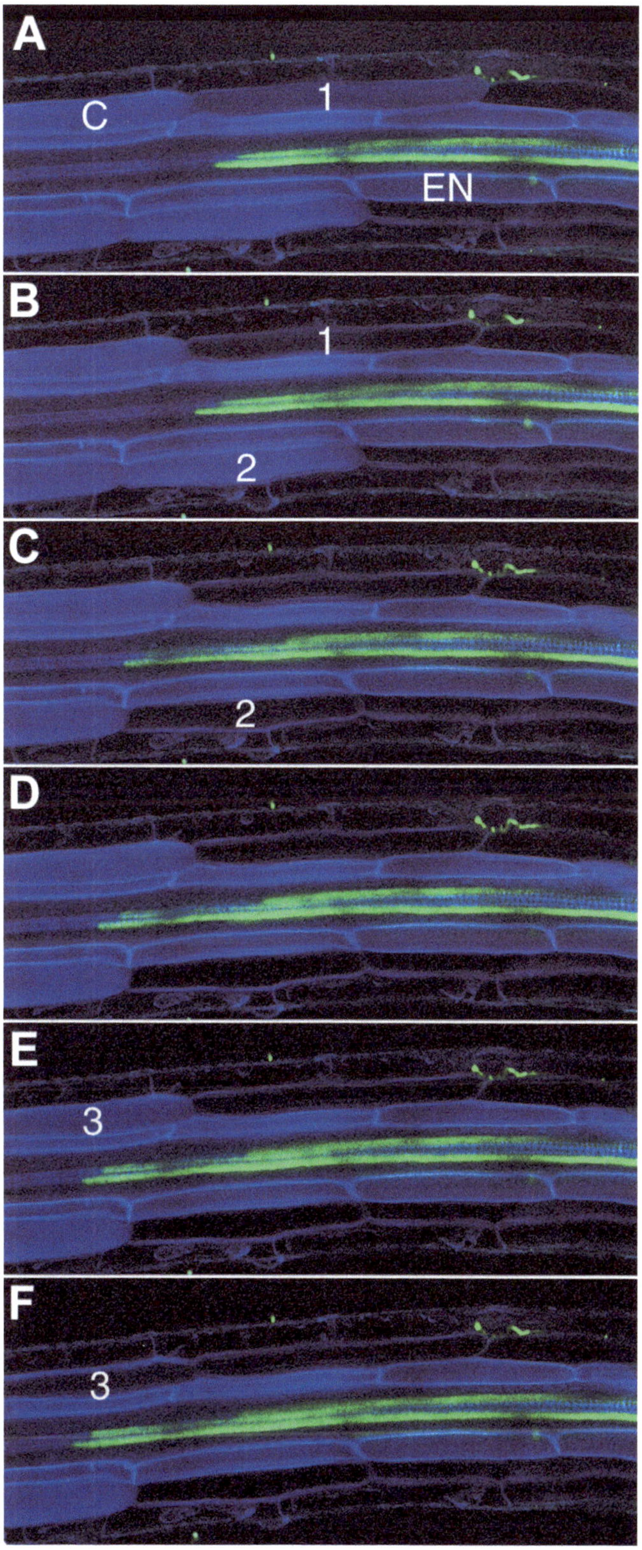

Fig. 5 Time-lapse microscopy of *F. oxysporum* growth within vascular bundle of *A. thaliana* Col-0 grown in agar-based medium. *F. oxysporum* progression through vascular tissue was documented using confocal microscopy of

bandpass (for autofluorescence of plant cell walls, vacuolar contents, and xylem elements) (Fig. 5) (*see* **Note 7**).

4. For 4-D datasets (3-D stacks over time), adjust the pinhole from 1 to 1.5 airy units and the z-spacing to match the optical slice thickness (*see* **Note 8**).
5. For optimal lateral resolution, acquire datasets with at least 1024 × 1024 pixel resolution. Greater pixel resolutions may be used to improve the lateral resolution if needed (objective lens dependent).
6. Use the LSM 510 AIM software (Rel. 3.2) MultiTime Macro (ver. Q) in conjunction with a motorized x-y stage to mark the coordinates and acquire z-stacks at multiple specified locations at 20–60 min intervals.
7. In the MultiTime Macro, use the auto-focus routine by capturing an x-z line-scan with the 633 laser line in reflection mode which will allow determination of the coverslip/media interface prior to image collection of each z-stack and location (*see* **Note 9**).
8. Save each z-stack as an independent dataset, immediately following acquisition in order to prevent data loss in instances of power and/or equipment failure and to manage the large amount of data generated from multiple locations over the duration of the experiment (typically 4–7 gigabytes/72 h period).
9. Concatenate individual 3-D z-stacks from each time point to create a 4-D dataset representing each location, and render each stack in the series as a maximum intensity projection.
10. Convert 4-D datasets into QuickTime or AVI movie files in which the rate of the interaction is viewed at an accelerated pace of ~10 frames/s (*see* **Note 10**).

3.5 Imaging Ca^{2+} Signature in Planta

We applied the imaging strategy for time-lapse in vivo visualization of *A. thaliana* roots colonized by *F. oxysporum* described above to characterize the dynamics of Ca^{2+}, a second messenger that regulates diverse processes in both plants and fungi, from both host and

Fig. 5 (continued) ZsGreen (green) and multiphoton microscopy of plant walls and cortex vacuolar contents (blue). As hyphae grew from right to left, blue vacuolar autofluorescence significantly decreased in cells adjacent to the leading hyphal tips. Numbers (1–3) in consecutive frames indicate the cortex cells (C) before and after loss of vacuolar fluorescence. Frames 1 (**a**), 2 (**b**), 7 (**c**), 9 (**d**), 10 (**e**), and 11 (**f**) of a 24 frame time-series (each frame captured at a 2 min 54 s time interval). Leading hypha growth rate 2.8 μm/min. EN (endodermis) C (cortex). Scale bar = 50 μm. (Reproduced with permission from Czymmek et al. [3])

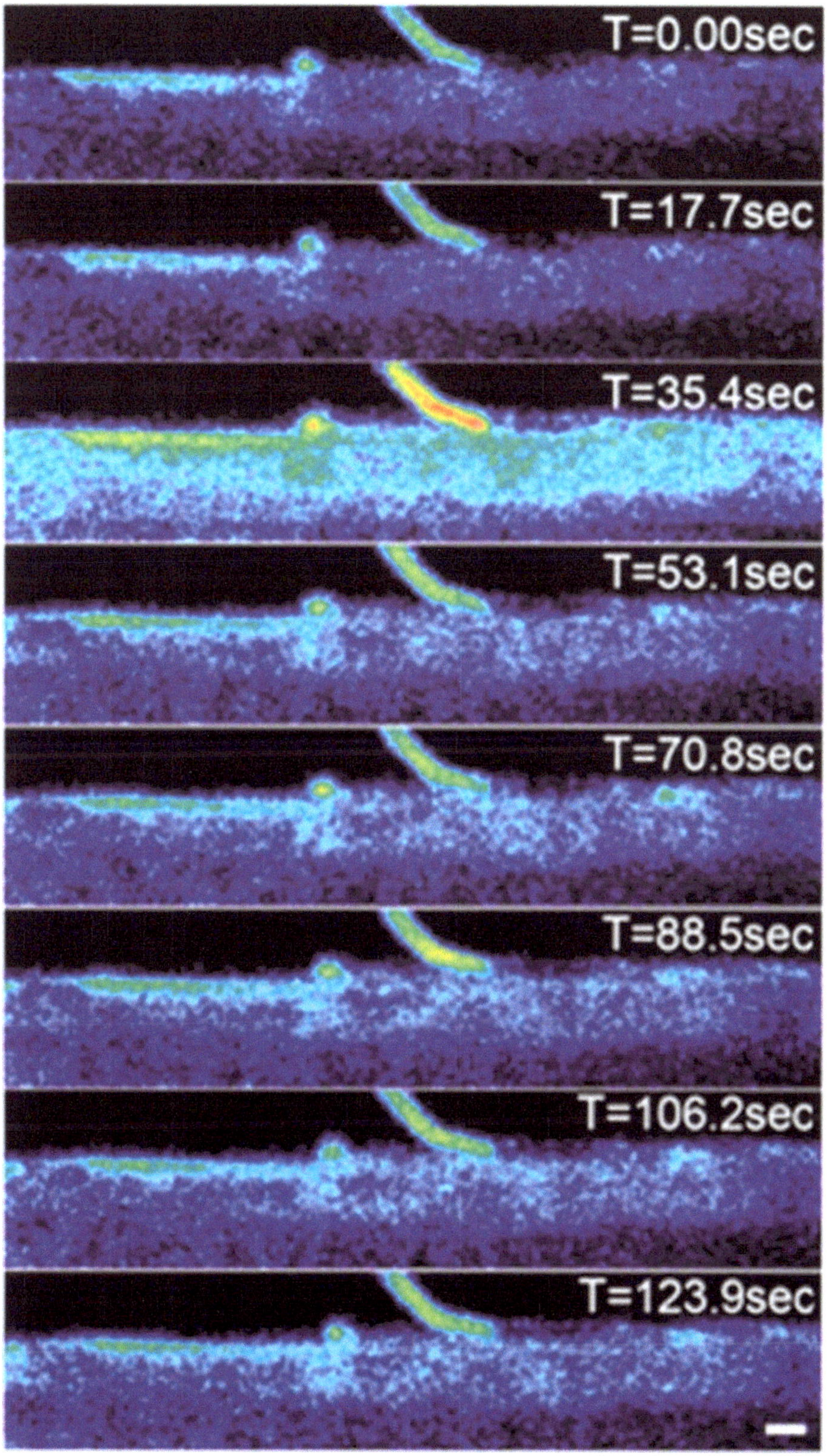

Fig. 6 Calcium response in both host and pathogen during *F. oxysporum*-*A. thaliana* interaction. This time-lapse sequence shows Ca^{2+} transients within seconds of contacting the root epidermis of YC 3.60 expressing *A. thaliana* ecotype Cvi-0 by *F. oxysporum*. A large Ca^{2+} transient ($T = 35.4$ s) followed by a smaller one

pathogen sides as they interact. To achieve this, we expressed YC3.60 [19] in both *A. thaliana* and *F. oxysporum* [10]. We were able to document cytoplasmic Ca^{2+} ($[Ca^{2+}]_c$) pulses in a hypha of *F. oxysporum* contacting the root surface of *A. thaliana* ecotype Cvi-0 expressing YC3.60 (Fig. 6) and hyphal progression through vascular tissue [10]. This approach clearly demonstrated the feasibility and utility of our imaging chamber system for tracking pathogenesis-related cell dynamics with FP fusions and biosensors. Slight modifications of our imaging protocol from above (*see* Subheading 3.3) to achieve this are described below.

1. Grow seedlings of *A. thaliana* ecotype Cvi-0 expressing YC3.60 in a single-welled Lab-Tek chambered coverglass system with 0.5× strength *Arabidopsis* nutrient agar (0.7%).
2. Perform time-lapse imaging on a Zeiss LSM 5 DUO laser scanning confocal microscope with a 40× oil Plan-NeoFluar (1.3 numerical aperture) objective lens.
3. Use the 458 nm laser line of a 25 mW argon ion laser with the laser power attenuated to prevent photobleaching.
4. Acquire images from three channels (YFP, CFP, and transmitted light), and include an online ratio image of YFP/CFP with background subtraction to more easily monitor $[Ca^{2+}]_c$ pulses in real-time (*see* **Note 11**).
5. For each fluorescent channel, set the pinholes for the YFP and CFP channels to its maximum (1000 μm diameter), and adjust the gain/offsets to yield an average ratiometric value of approximately 1 in the cell cytoplasm.
6. Acquire each z-stack at multiple locations, and process into time-lapse movies as described above (*see* Subheading 3.4).

4 Notes

1. Do not supplement media for soil and agar-based imaging analyses with sucrose as it results in *F. oxysporum* to cause a severe infection, regardless of the level of *F. oxysporum* resistance among the tested ecotypes.
2. Prepare individual stock solutions separately and filter sterilize to prevent precipitation. For 1 L Hoagland solution, add

Fig. 6 (continued) ($T = 88.5$ s) in *F. oxysporum*, while in *A. thaliana*, a cytoplasmic Ca^{2+} response coincided with *F. oxysporum* contact and was elevated for the remainder of the sequence. Scale bar = 10 μm. (Reproduced with permission from Kim et al. [10])

sterilized solutions sequentially to 986.5 mL of autoclaved ddH_2O after cooling down to room temperature.

3. Equivalent manufacturer with compatible multi-dimensional hardware and software can be substituted.
4. When a large number of spores are needed, we cultured the transformants in CMC broth.
5. The Lab-Tek chambers can be periodically removed from the incubator and surveyed via fluorescence and/or brightfield microscopy (agar assay) or fluorescence microscopy only (soil assay) on an inverted microscope to assess fungal progression toward the roots.
6. Strict use of high magnification lenses often resulted in failure to document primary penetration events due to limited field-of-view.
7. Laser powers should be reduced as described previously [2] to prevent photobleaching, phototoxic effects, and other light-induced artifacts.
8. A trade off in z-resolution in the final 3D dataset often is required to match the experimental kinetics of the biological process. Under-sampling by matching the z-step to the optical slice thickness is a reasonable compromise to capture dynamic events through a tissue volume and minimize exposure to laser light.
9. Due to thermal and other motion considerations, z-drift during time-lapse imaging can be problematic. This can be addressed by an appropriate auto-focus routine.
10. For qualitative visualization purposes, a gamma correction can be applied equally to all images of certain time-lapse sequences to enhance visualization of low-intensity fluorescence.

    Application of a gamma correction must be reported in published methods.
11. For ratiometric Ca^{2+} imaging, use a 465–510 nm bandpass and 530 nm longpass emission filters for CFP and YFP FRET signals, respectively.

Acknowledgments

Most of the materials and associated protocols described here were developed using support from USDA-NRI (2002-02367). Generation of *F. oxysporum* transformants expressing YC3.60 and analyses of their Ca^{2+} signatures as they encounter roots and proliferate in the root vascular system were supported by the US National Science Foundation (MCB-1051667). The USDA Specialty Crop Multi-State Program (AM170200XXXXG006), the USDA

National Institute of Food and Agriculture and Federal Appropriations (Project PEN04655; Accession # 1016291), and the US Department of Agriculture, Agricultural Research Service, supported the preparation of this chapter.

References

1. Duniway JM (2002) Status of chemical alternatives to methyl bromide for pre-plant fumigation of soil. Phytopathology 92:1337–1343
2. Czymmek KJ, Bourett TM, Sweigard JA et al (2002) Utility of cytoplasmic fluorescent proteins for live-cell imaging of *Magnaporthe grisea* in planta. Mycologia 94:280–289
3. Czymmek KJ, Fogg M, Powell DH et al (2007) *In vivo* time-lapse documentation using confocal and multi-photon microscopy reveals the mechanisms of invasion into the Arabidopsis root vascular system by *Fusarium oxysporum*. Fungal Genet Biol 44:1011–1023
4. Kim H-S, Park S-Y, Lee S et al (2011) Loss of cAMP-dependent protein kinase a affects multiple traits important for root pathogenesis by *Fusarium oxysporum*. Mol Plant-Microbe Interact 24:719–732
5. Nishimura MT, Dangl JL (2010) Arabidopsis and the plant immune system. Plant J 61:1053–1066
6. Woodward AW, Bartel B (2018) Biology in bloom: a primer on the *Arabidopsis thaliana* model system. Genetics 208:1337–1349
7. Agrawal S (2018) *Arabidopsis thaliana* as a model organism to study plant-pathogen interactions. In: Molecular aspects of plant-pathogen interaction. Springer Singapore, Singapore, pp 1–20
8. Dini-Andreote F, Raaijmakers JM (2018) Embracing community ecology in plant microbiome research. Trends Plant Sci 23:467–469
9. Pascale A, Proietti S, Pantelides IS et al (2020) Modulation of the root microbiome by plant molecules: the basis for targeted disease suppression and plant growth promotion. Front Plant Sci 10:1741
10. Kim H-S, Czymmek KJ, Patel A et al (2012) Expression of the Cameleon calcium biosensor in fungi reveals distinct Ca^{2+} signatures associated with polarized growth, development, and pathogenesis. Fungal Genet Biol 49:589–601
11. Bourett TM, Sweigard JA, Czymmek KJ et al (2002) Reef coral fluorescent proteins for visualizing fungal pathogens. Fungal Genet Biol 37:211–220
12. Monshausen GB, Messerli MA, Gilroy S (2008) Imaging of the yellow Cameleon 3.6 indicator reveals that elevations in cytosolic Ca^{2+} follow oscillating increases in growth in root hairs of Arabidopsis. Plant Physiol 147:1690–1698
13. Murashige T, Skoog F (1962) A revised medium for rapid growth and bio assays with tobacco tissue cultures. Physiol Plant 15:473–497
14. Akerboom J, Calderón NC, Tian L et al (2013) Genetically encoded calcium indicators for multi-color neural activity imaging and combination with optogenetics. Front Mol Neurosci 6:1–29
15. Mullins ED, Chen X, Romaine P et al (2001) Agrobacterium-mediated transformation of Fusarium oxysporum: an efficient tool for insertional mutagenesis and gene transfer. Phytopathology 91:173–180
16. Khang CH, Park S-Y, Rho H-S et al (2006) Filamentous Fungi (*Magnaporthe grisea* and *Fusarium oxysporum*). In: Agrobacterium Protocols Volume 2. Humana Press, Totowa, NJ, pp 403–420
17. Proctor RH, Desjardins AE, Plattner RD et al (1999) A polyketide synthase gene required for biosynthesis of fumonisin mycotoxins in *Gibberella fujikuroi* mating population a. Fungal Genet Biol 27:100–112
18. Clough SJ, Bent AF (1998) Floral dip: a simplified method for agrobacterium-mediated transformation of *Arabidopsis thaliana*. Plant J 16:735–743
19. Nagai T, Yamada S, Tominaga T et al (2004) Expanded dynamic range of fluorescent indicators for Ca^{2+} by circularly permuted yellow fluorescent proteins. Proc Natl Acad Sci U S A 101:10554–10559

Chapter 14

High-Throughput Screening Assays to Identify Plant Natural Products with Antifungal Properties Against *Fusarium oxysporum*

Yong Zhang, Pei-Lun Kao, Akaansha Rampal, Sibongile Mafu, Sergey Savinov, and Li-Jun Ma

Abstract

Fusarium oxysporum is a cross-kingdom fungal pathogen that not only causes devastating plant vascular diseases but can also opportunistically infect humans. Here we describe two high-throughput screening assays, a resazurin cell viability assay and an optical density assay, to screen natural products from cultured plant cells with antifungal properties against a clinical isolate of *F. oxysporum*. After elicitation by applying methyl jasmonate or by co-culture with *F. oxysporum*, as an abiotic elicitor and a biotic elicitor, respectively, we identified three cell lines that produce materials that inhibit fungal growth. Our procedure validates the powerful potential of combining high-throughput methods for the discovery of novel anti-pathogenic leads.

Key words *Fusarium oxysporum*, High-throughput, Assay development, Antifungal screen, Plant cell culture collection

1 Introduction

Fusarium oxysporum is a cross-kingdom fungal pathogen that is widely distributed in diverse environments, including soil, indoors, and aquatic habitats [1]. Members of *F. oxysporum* not only cause devastating plant vascular diseases but also infect immunocompromised individuals, resulting in localized or even life-threatening disseminated infections [2, 3]. Plants are known to produce a large repertoire of specialized metabolites [4, 5], offering great potential in discovering new drugs with distinct modes of action for developing new antifungal drugs [6, 7]. The Plant Cell Culture Library (PCCL) at UMass Amherst contains over thousand plant cell cultures representing ecologically and taxonomically diverse plant species, offering a rich resource of phytochemicals for the

Jeffrey Coleman (ed.), *Fusarium wilt: Methods and Protocols*,
Methods in Molecular Biology, vol. 2391, https://doi.org/10.1007/978-1-0716-1795-3_14,

discovery of novel compounds with distinct modes of action to control fungal diseases.

In this chapter, we present protocols established for screening plant cell extracts to identify those with antifungal activity, including high-throughput screening assays to evaluate effectiveness of potential antifungals and for obtaining extracts from plant cell lines. We report results of evaluating these protocols against *F. oxysporum* NRRL 32931, a clinical isolate responsible for a disseminated fusariosis [3].

2 Materials

1. Fungal Strain: *F. oxysporum* NRRL 32931.
2. Murashige–Skoog (MS) medium: Add 3 g Gamborg's B5 salts, 30 g sucrose, 4.33 g MS salts, and 15 g agar to 800 mL ddH_2O, and stir until dissolved. Adjust pH to 6.4 and bring to a final volume of 1 L with dd H_2O. Sterilize by autoclaving (*see* **Note 1**).
3. Chemicals

   Antifungals: the stock solutions of each commercial antifungal were prepared by dissolving weighted material in dimethyl sulfoxide (DMSO).

   (a) Amphotericin B stock solution (20 mg/mL): Dissolve 200 mg amphotericin B in 10 mL DMSO, and sterilize by filtration and store at −20 °C.

   (b) Fluconazole stock solution (100 mg/mL): Dissolve 1 g fluconazole in 10 mL DMSO, and sterilize by filtration and store at −20 °C.

   (c) Nystatin stock solution (3.2 mg/mL): Dissolve 32 mg nystatin in 10 mL DMSO, and sterilize by filtration and store at −20 °C.

   (d) Phenamacril stock solution (0.4 mg/mL): Dissolve 4 mg phenamacril in 10 mL DMSO, and sterilize by filtration and store at −20 °C.

   (e) Resazurin solution (0.02%, w/v): Dissolve 0.02 g resazurin sodium salt in 100 mL of sterile distilled water, and sterilize by filtration. All the experiments are performed using 0.002% (w/v) as the final concentration of the dye (*see* **Note 2**).

   (f) 1 mM methyl jasmonate (MJ) elicitation medium: Add 2.22 μL of MJ to 9.998 mL MS liquid medium.
4. Plant Cell Cultures (https://www.umass.edu/ials/pccl-database).

5. Consumables and Other Supplies.
   (a) 96-well microplates.
   (b) Spectrophotometer for 96-well plates.
   (c) Orbital shaker for plates.
   (d) Light box.
   (e) Microcentrifuge tubes.
   (f) Methanol.
   (g) 2 mm metal beads.
   (h) Tissue homogenizer.
   (i) Vacuum centrifuge.
   (j) Vortex.
   (k) DMSO-compatible sealing foil for microplates.
   (l) Bi-chamber transwell plates (with 0.4 μm polycarbonate membranes) (Corning Inc., Corning, NY).

3 Methods

3.1 Cultivation and Preparation of Fungi

1. Inoculate a MS agar plate with a thawed glycerol stock of *F. oxysporum* from −80 °C freezer on a MS agar plate and incubated at 28 °C for 3 days.
2. Transfer a small piece of the fungal agar into MS liquid media, and place on shaker at 28 °C for 150 rpm to grow biomass.
3. Collect spore suspension by filtering the fungal culture through three layers of miracloth. Concentration of fungal spores is determined by counting the conidia in a hemocytometer counting chamber. All the following experiments are performed using 1×10^6 spores/mL as the final concentration of fungi (*see* **Note 3**).

3.2 Developing High-Throughput Screening Assays

We have adapted and validated two high-throughput screening assays, a resazurin cell viability assay, and an optical density assay both of which assess fungal cell viability and activities to investigate antifungal activities against *F. oxysporum*. Both methods are simple, straightforward, and capable of producing rapid results at a low cost. Four commercial antifungals, amphotericin B (AmB), fluconazole (Flu), nystatin (Nys), and phenamacril (Phe), were used to standardize the assay conditions in positive control wells. Each 96-well plate always includes positive control wells, negative control wells, and treatment wells with replicates. The negative control includes 178 μL MS, 20 μL of 1×10^7 spores/mL *F. oxysporum* in MS and 2 μL of DMSO (*see* **Note 4**).

3.2.1 Resazurin Cell Viability Assay

Resazurin, a water-soluble, non-toxic, non-fluorescent blue dye, has been used as a redox indicator of metabolic activities and viability of cells. This feature enabled us to develop a robust screening assay to assess antifungal property of plant-derived materials against *F. oxysporum*.

Figure 1 shows two replicates of four commercial antifungal treatment wells. Each well contains 178 μL of MS, 20 μL of 1×10^7 spores/mL *F. oxysporum* in MS, 2 μL of 100× antifungal at each concentration (1% v/v) to make the following concentrations: a) Amphotericin B (AmB): 0.064, 0.32, 1.6, 8, 40, 200 μg/mL; b) Fluconazole (Flu): 0.32, 1.6, 8, 40, 200, 1000 μg/mL; c) Nystatin (Nys): 1, 2, 4, 8, 16, 32 μg/mL; and d) Phenamacril (Phe): 0.125, 0.25, 0.5, 1, 2, 4 μg/mL.

The top wells without antifungals are negative controls. The bottom wells are positive controls that lack fungal spores and contain the highest concentrations of antifungal solutions (AmB, 200 μg/mL; Flu, 1000 μg/mL; Nys, 32 μg/mL; Phe, 4 μg/mL) (*see* **Note 5**).

The protocol includes the following steps:

1. Incubate the above 96-well plates for 16 h at room temperature with gentile agitation (150 rpm on an orbital shaker).
2. Add ~20 μL of the 0.008% resazurin solution into each well.
3. Wrap the plate with aluminum foil, and incubate on a shaker at 150 rpm for 20 min.
4. Observe the extent of blue resazurin reduction to pink resorufin on the light box (Fig. 1), and make a photographic record.

3.2.2 Optical Density Assay

Monitoring growth using a spectrophotometer is one of the simplest methods to measure microbial growth by tracking changes in the optical density (OD) over time. Critical parameters for the success of this assay include initial spore concentration, culture medium, temperature, and sampling time points. Figure 2 shows three replicates of four commercial fungicide treatments. Each well contains 178 μL MS, 20 μL of 1×10^7 spores/mL *F. oxysporum* in MS, 2 μL of 100× antifungal at each concentration (1%, v/v) to make the screening concentration as: a) Amphotericin B (AmB): 0.064, 0.32, 1.6, 8, 40, 200 μg/mL; b) Fluconazole (Flu): 0.32, 1.6, 8, 40, 200, 1000 μg/mL; c) Nystatin (Nys): 1, 2, 4, 8, 16, 32 μg/mL; d) Phenamacril (Phe): 0.125, 0.25, 0.5, 1, 2, 4 μg/mL.

The protocol includes the following steps:

1. Place the 96-well plate from Subheading 3.2.1 in the spectrophotometer.
2. Set the plate reader to read optical density at 600 nm, at intervals of 1 h.

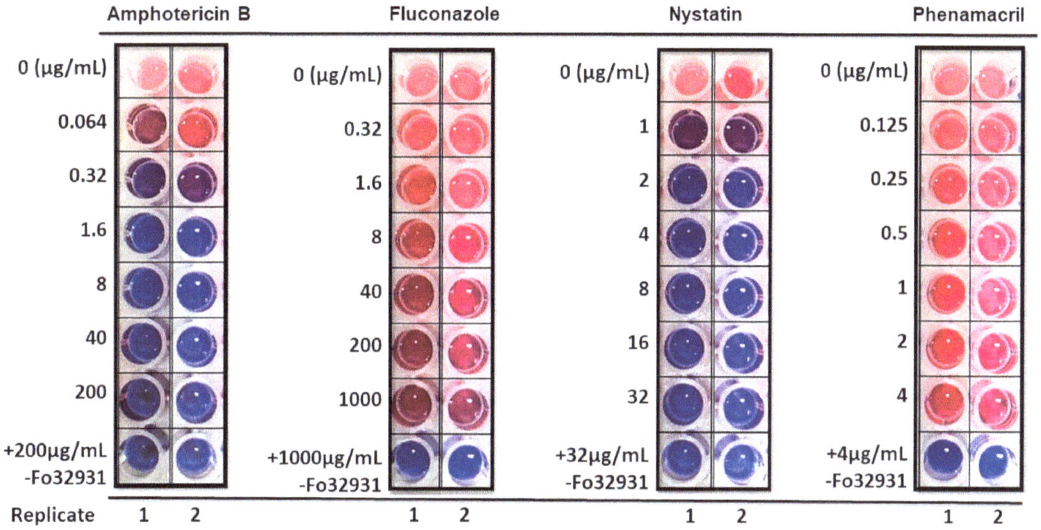

Fig. 1 Resazurin cell viability assay. After 16-h inoculation, resazurin was added to the assay plate. The image was taken after a 20-min incubation with shaking

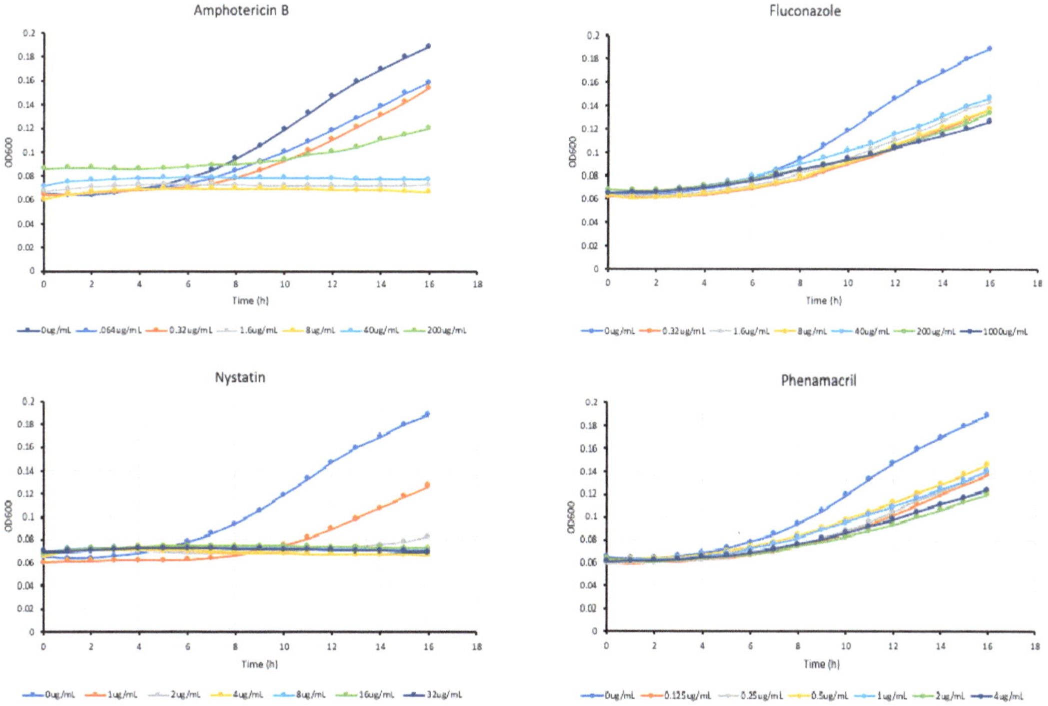

Fig. 2 Results of optical density assay. Growth curves (optical density, OD = 600 nm) of *F. oxysporum* on different concentrations of commercial fungicides. Each value was calculated using mean of triplicates for each testing

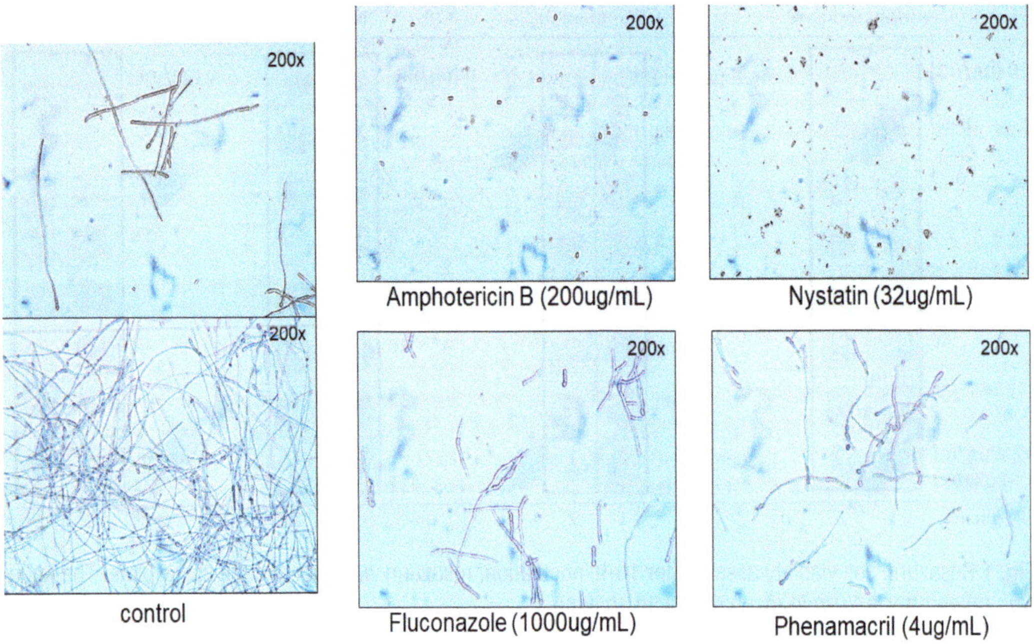

Fig. 3 Morphological validation. *F. oxysporum* treated with fungicides: 16-h post inoculation, at 200× magnification of light microscope
Control: *F. oxysporum* without antifungal treatment

3. Set temperature control to 25 °C.
4. Incubate without shaking for 16 h.

Results from both the resazurin cell viability and the optical density assays are confirmed using microscopic observation of *F. oxysporum* viability and potential growth (Fig. 3). Briefly, take 50 μL of the fungal culture under the same treatment as described above from each well onto a glass slide; before loading, mix each well thoroughly to ensure an evenly distributed mixture.

3.3 Screening Plant Cell Lines for Compounds with Antifungal Properties

In general, plant secondary metabolites are produced at a relatively low level, if any, in cultured cell lines, as primary metabolite precursors are primarily used for biomass accumulation. One of the most effective strategies for enhancing the production of secondary metabolites is elicitation, using agents that act as a signal to stimulate the biosynthesis and accumulation of secondary metabolites [8–10]. We tested the induction of plant secondary metabolites with potential antifungal properties against *F. oxysporum* by applying methyl jasmonate or live *F. oxysporum* as the abiotic and biotic elicitors, respectively (*see* **Note 6**).

3.3.1 Abiotic Elicitation Using Methyl Jasmonate

Here we report the results of treating randomly selected PCCL cell lines by methyl jasmonate (MJ) (*see* **Note 7**).

Preparation of MJ-elicited Plant Callus

1. Transfer 100 mg of plant callus culture into a 2 mL microfuge tube in duplicate (two microfuge tubes per species).
2. Add 1 mL of MJ elicitation medium into one tube and 1 mL MS medium into the other tube.
3. Incubate samples on shaker at 150 rpm for 24 h at room temperature.

Preparation of Cellular Extracts

4. Centrifuge all tubes at 5000 × *g* for 10 min and remove the supernatant.
5. Wash the plant cells by adding 1 mL sterile water into elicited/non-elicited tube, and agitate by gentle vortexing of the tubes to break up the cell pellet.
6. Spin down the samples at 1000 rpm for 10 min and discard the washings. Repeat the centrifugation step, if necessary, to remove as much liquid as possible.
7. Combine the resulting cell pellets by adding 1 mL of methanol and two 2 mm ball bearings to each tube. Homogenize the sample in for 1.5 min at 25 Hz.
8. Spin down the samples at 12,000 rpm for 10 min, and transfer the supernatant into a pre-tared microfuge tube.
9. Evaporate all volatile components in a vacuum centrifuge.
10. Record the weight of residues and re-dissolve in DMSO to a concentration of 20 mg/mL. Vortex and/or sonicate to ensure complete solubilization.
11. Centrifuge or filter the resulting solution and transfer to wells of a cryo-compatible multiwell plate; cover the plate with DMSO-compatible sealing foil and store at −20 °C.

Preparing the Plate for Screening

12. Mix negative control wells containing 20 μL 1 × 10^7 spores/mL of *F. oxysporum* spores in MS, 4 μL DMSO, and 176 μL MS media.
13. Mix positive control wells containing 20 μL 1 × 10^7 spores/mL of *F. oxysporum* spores in MS, 4 μL 400 μg/mL nystatin, and 176 μL MS media.
14. For the treatment wells of the extract(s) to be tested, mix 4 μL of MJ-elicited /non-elicited callus extracts, 20 μL 1 × 10^7 spores/mL of *F. oxysporum* spores in MS, and 176 μL of MS media.
15. After incubation for 16 h at room temperature, ~20 μL of resazurin solution is applied to each well, and the plate is imaged after a 20 min incubation at room temperature.

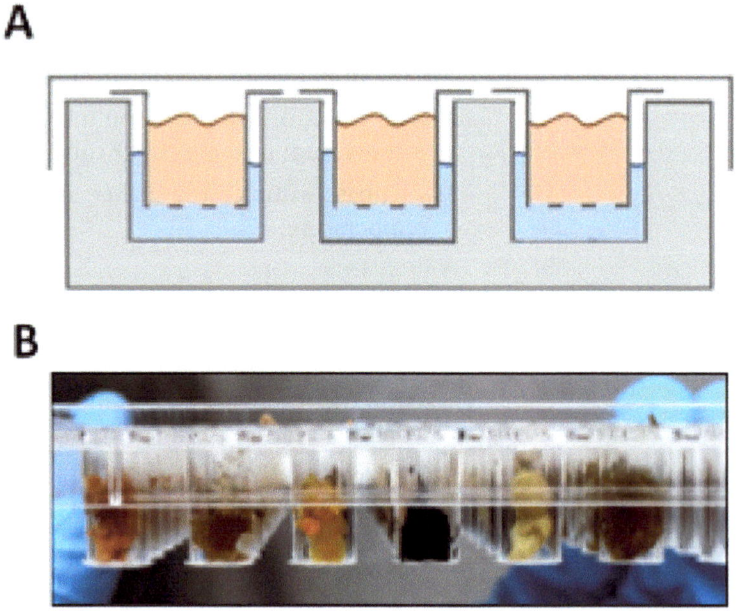

Fig. 4 The transwell system. (**a**) A schematic representation of suspension of microbial cells (blue) in a microwell plate (grey) is combined with an insert containing a porous (0.4 μm) membrane at the bottom of each well (dashed lines). (**b**) Aliquots of plant cultures are suspended in medium within the insert wells (beige)

3.3.2 Biotic Elicitation Using F. oxysporum *in a Bi-chamber Transwell Plate*

Here we describe a high-throughput screening method for antifungal plant cell extracts using 96-well bi-chamber transwell plates. In this format, a plant cell culture and a biotic elicitor can be separated by a porous membrane. The pore size ranges from 0.4 to 8.0 μm in diameter for variable permeability. The pores on the membrane are small enough to prevent the physical cell to cell contact across the membrane, but allow for signaling molecules secreted from either cell cultures to diffuse through the membrane enabling bi-directional exchange between chambers (*see* **Notes 7** and **8**).

In our experimental setup (Fig. 4), the plant cell cultures are loaded on the top chamber of the transwell plate, and spores of *F. oxysporum* NRRL 32931 are added to the bottom chamber, with both wells suspended in MS liquid medium. This setup permits diffusion of fungal elicitors, their interaction with the plant cells, and induction of biosynthesis and release of phytochemical compounds with antifungal properties. The antifungal compounds produced by plant cell can be extracted from bottom wells and subjected to metabolomics studies to identify molecules responsible for the antifungal activities.

Loading Plant Cells in Upper Wells

1. Retrieve plant cell lines from the PCCL.
2. Harvest 0.05 ± 0.005 g of biomass (in three replicates) from each PCCL cell line culture plate, and load into the upper wells of a bi-chamber transwell plate (0.4 μm polycarbonate membrane) in a biosafety cabinet.
3. Each plant cell line has three replicates.

Loading Treatments in Lower Wells

4. Set up negative control wells containing 20 μL 1×10^7 spores/mL of *F. oxysporum* in MS, 4 μL DMSO, and 176 μL MS media.
5. The positive control wells are mixed by adding 20 μL 1×10^7 spores/mL of *F. oxysporum* in MS, 4 μL 400 μg/mL nystatin in DMSO, and 176 μL MS media.
6. For assessment of each tested cell line, add 4 μL DMSO, with 20 μL 1×10^7 spores/mL of *F. oxysporum* in MS and 76 μL of MS media to the cell line well.

Co-culture Incubation

7. The preloaded upper wells are inserted into the plate on top of the bottom wells, and 100 μL of MS media is added onto the upper sample wells to fully submerge the cells (Fig. 4).
8. Following the addition of *F. oxysporum* and the MS media, the transwell plate is covered by the lid, wrapped in aluminum foil, and placed on an orbital shaker set at 100 rpm, for 24 h at room temperature.

Evaluate Antifungal Activity

9. After 24 h of co-culture incubation, upper wells containing the plant cells are separated from the bottom wells in a biosafety cabinet.
10. Then ~20 μL of resazurin solution is applied to each of the bottom wells, and color changes are imaged (*see* **Note 8**).

4 Notes

1. Many available media can be used for growing fungi. However, potato dextrose agar (PDA) medium is too rich to be used in high-throughput screening assays.
2. Due to resazurin being light-sensitive, the resazurin solution should be wrapped in aluminum foil and stored at 4 °C. The resazurin solution is usually made as a 10× stock solution, and diluted before use. To make the experiment efficient and

accurate, we suggest preparing a fresh resazurin solution every month.

3. Before loading the sample onto hemacytometer, mix spore suspension thoroughly, and obtain an evenly distributed mixture. Fungal spore suspension is usually made as 1×10^7 spores/mL, and diluted to a final concentration of 1×10^6 spores/mL in the assays.
4. A concentration gradient has been tested to show the effect of DMSO on fungal growth. DMSO (v/v, $\leq$4%) does not affect fungal growth. The fungal inhibition in the experiment is, therefore, directly caused by the antifungal compounds.
5. Sometimes resazurin changes color due to the reaction with compounds in solution (i.e., antioxidants). Adding compounds with resazurin without any *F. oxysporum* as control can be used to check for a color change due to assay compounds. If resazurin changes to resorufin (pink) without *F. oxysporum*, this indicates that the compounds interfere with resazurin, giving false positive results. Hence, it is beneficial to test the compounds separately with resazurin as a control. The optical density assay could be used for such samples.
6. Our preliminary results demonstrate both the validity of the high-throughput screening methods and the great potential of plant natural products in identifying novel antifungals. Both MJ and bi-chamber transwell co-culture tactics were effective in eliciting plant cells to produce phytochemicals with antifungal properties. Such discoveries are much needed to manage the alarming combination of rapidly emerging multidrug-resistant microorganisms, an increasing population of immunocompromised individuals, and cross-border travel in the globalized world, whereby previously manageable infectious diseases could get out of control and become pandemic [11, 12].
7. The consistency of resazurin cell viability assay and optical density assay confirms that these rapid, simple, and robust high-throughput assays can be used to screen for antifungal activities using both commercially available small molecular libraries and novel natural products. Revealed by the MJ elicitation, 10 out of 40 plant cell lines produced a certain level of inhibition of fungal growth. The extracts from MJ-elicited cells showed growth inhibition, as detected by resazurin processing in sample and control wells, with some of them approaching the effect seen in positive control (8 μg/mL nystatin) wells, suggesting that plant secondary metabolite biosynthesis genes can be induced by subjecting plant cells to stress, which are able to produce anti-fusarial compounds (Fig. 5). We also tested 14 plant cell lines using the transwell assay by co-culturing

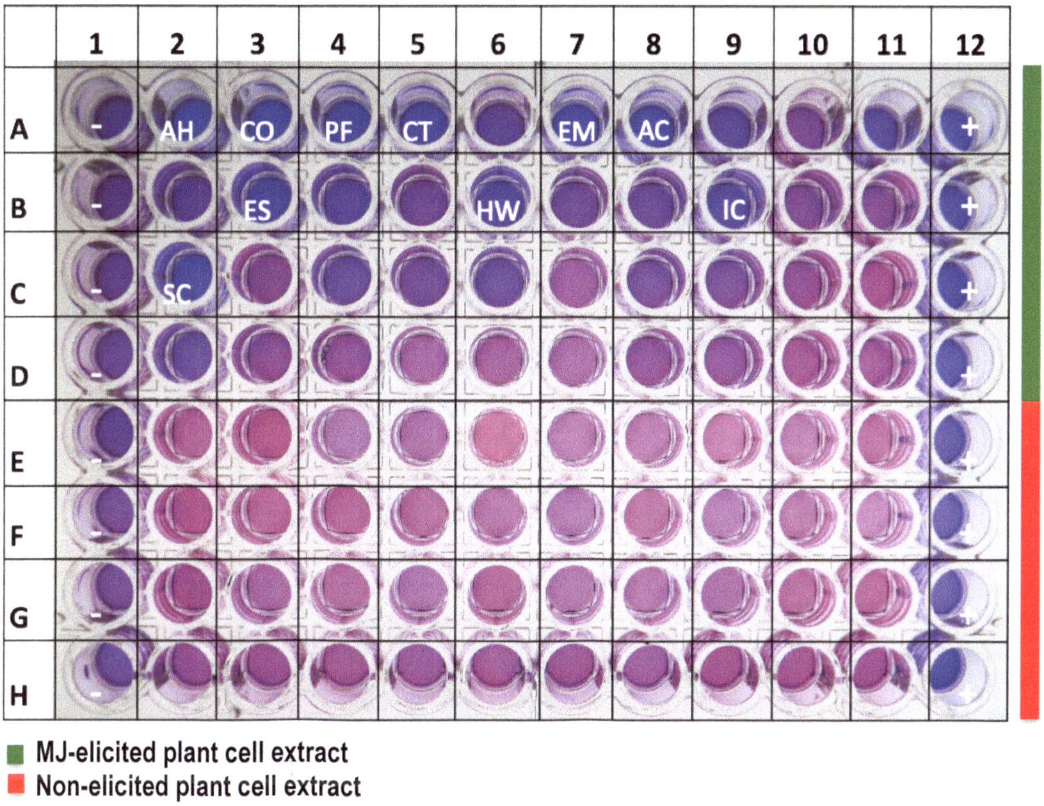

Fig. 5 Results after abiotic elicitation using methyl jasmonate. High-throughput screening of antifungal drug activity using 40 plant cell lines selected from the PCCL. The first and the last columns are negative (−) and positive (+) controls, respectively; wells in the first four rows of columns 2–11 were loaded with MJ-elicited extracts; wells in the bottom four rows of columns 2–11 were loaded with extracts from the corresponding cell lines without the elicitation. Ten extracts with antifungal property are highlighted with plant cell line IDs: AH, *Allocasuarina helmsii*; CO, *Catalpa ovata*; PF, *Peperomia fraseri*; CT, *Costus tappenbeckianus*; EM, *Enterolobium multiflorum*; AC, *Agrostis castellana*; ES, *Eucalyptus saligna*; HW, *Hadrodemas warszewiczianum*; IC, *Ipomoea costata*; SC, *Strychnos colubrine*

these plant cells with *F. oxysporum* in 96-well transwell plates. Five hits were identified with anti-fusarial activities (Fig. 6). Interestingly, three of them—IC, *Ipomoea costata*; EM, *Enterolobium multiflorum*; and SC, *Strychnos colubrine*—are identified by both assays. The former two, IC and EM, have displayed the highest anti-fusarial activities, as detected by the resazurin viability assay. These three plant species are reported for the first time to possess components with antifungal properties, while the families to which the three plants belong do contain species that produce antifungal compounds [13–15].

8. In contrast to other elicitation methods, in situ co-culture elicitation is more easily operated than cell suspension cultures because the plant cells are self-immobilized and retained within

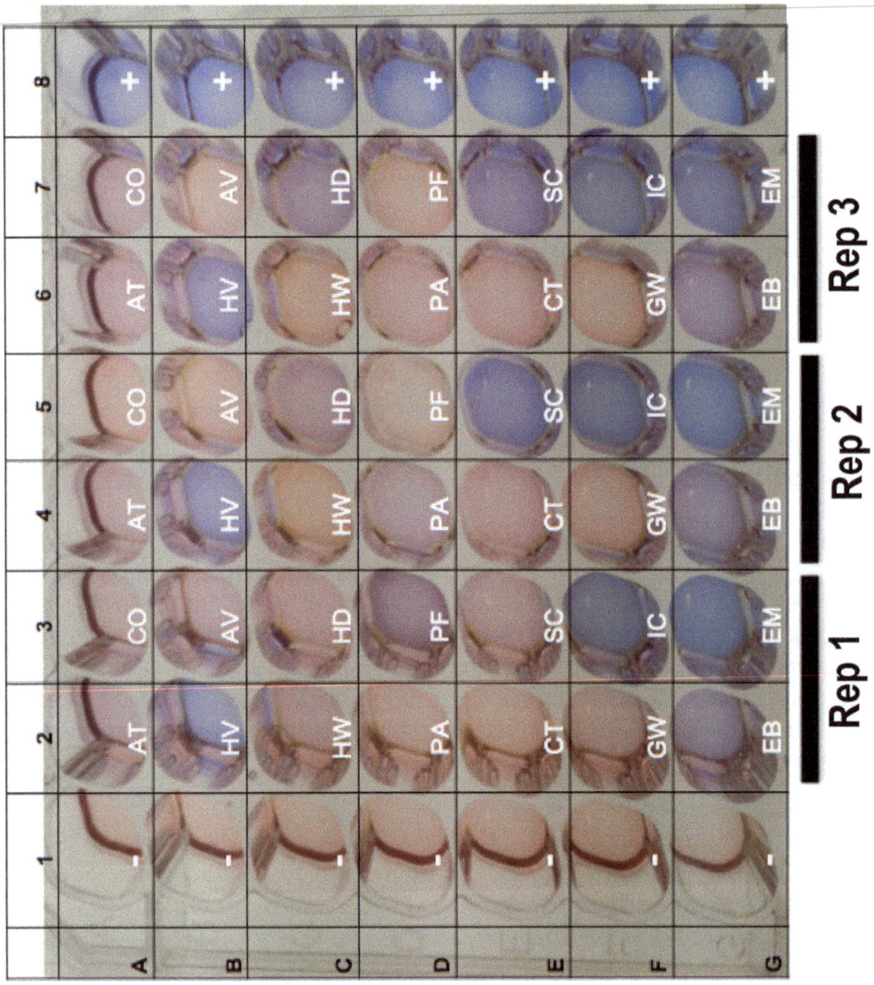

Fig. 6 Antifungal screening using transwell co-culture technique. The first column is negative control without antifungals; the last column is positive control with 8 μg/mL nystatin. Each plant species has three replicates. Five hits were found with anti-fungal activities. The strength of the plant species with anti-fusarial activities can be ranked as: IC > EM > SC > HV > EB. IC, *Ipomoea costata*; EM, *Enterolobium multiflorum*; SC, *Strychnos colubrine*; HV, *Hordeum vulgare*; EB, *Eucalyptus bosistoana*

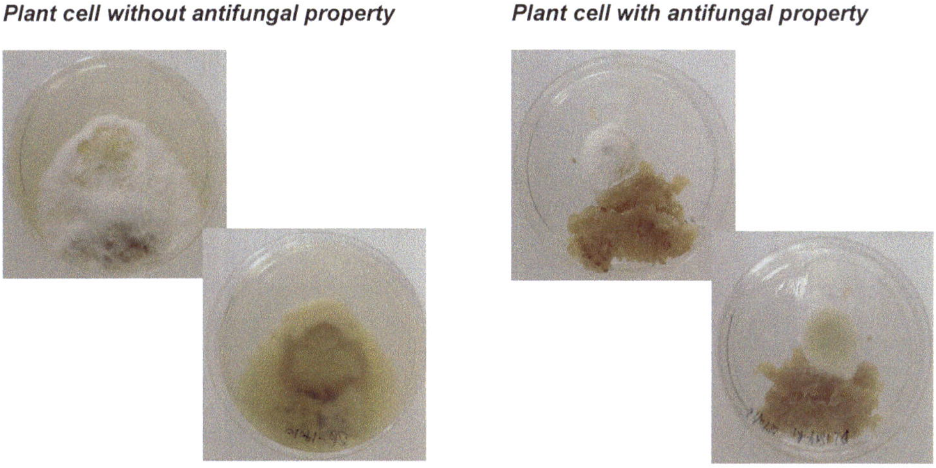

Fig. 7 In situ co-culture elicitation. *F. oxysporum* strain NRRL 32931 is inoculated in the center of each plate, and different plant cell lines are on the side, showing both top and bottom views

the culture plate. The antifungal activity is very straightforward, as indicated in Fig. 7. The exact incubation time of resazurin and the assay solutions is dependent on the color change of the negative control. The optimal time is when the initially blue resazurin is fully converted to pink resorufin.

Acknowledgments

We thank Katie Webster, Michael Daley, and many undergraduate students for their dedication in maintaining the PCCL cell lines; UMass Institute for Applied Life Sciences Core Facilities for maintaining the PCCL; Jennifer Normanly and Elizabeth Vierling for their consistent leadership in securing various funds to support the PCCL and related activities; and Jennifer Normanly for her involvement in the development of this research and her critical review of this manuscript.

Funding: The establishment of the PCCL has been sponsored by National Science Foundation (NSF/CSBR-1561572), US Department of Agriculture Massachusetts Experiment Station Awards (MAS00496 and MAS00520), the Science and Technology fund from the President's Office of the University of Massachusetts, and the 2016 UMass Amherst Armstrong Fund for Science. YZ and LJM are supported by the National Institute of Health (R01EY030150) and Burroughs Welcome Foundation (1014893).

References

1. Ma LJ, Geiser DM, Proctor RH, Rooney AP, O'Donnell K, Trail F et al (2013) Fusarium pathogenomics. Annu Rev Microbiol 67:399–416
2. Zhang Y, Ma LJ (2017) Deciphering pathogenicity of Fusarium oxysporum from a Phylogenomics perspective. Adv Genet 100:179–209
3. Zhang Y, Yang H, Turra D, Zhou S, Ayhan DH, DeIulio GA et al (2020) The genome of opportunistic fungal pathogen Fusarium oxysporum carries a unique set of lineage-specific chromosomes. Commun Biol 3:50
4. Rao SR, Ravishankar GA (2002) Plant cell cultures: chemical factories of secondary metabolites. Biotechnol Adv 20:101–153
5. Dixon RA (2001) Natural products and plant disease resistance. Nature 411:843–847
6. Banthorpe DV (1994) Secondary metabolism in plant tissue culture: scope and limitations. Nat Prod Rep 11:303–328
7. Zubricka D, Misianikova A, Henzelyova J, Valletta A, De Angelis G, D'Auria FD et al (2015) Xanthones from roots, hairy roots and cell suspension cultures of selected Hypericum species and their antifungal activity against Candida albicans. Plant Cell Rep 34:1953–1962
8. Akula R, Ravishankar GA (2011) Influence of abiotic stress signals on secondary metabolites in plants. Plant Signal Behav 6:1720–1731
9. Rejeb IB, Pastor V, Mauch-Mani B (2014) Plant responses to simultaneous biotic and abiotic stress: molecular mechanisms. Plan Theory 3:458–475
10. Narayani M, Srivastava S (2017) Elicitation: a stimulation of stress in in vitro plant cell/tissue cultures for enhancement of secondary metabolite production. Phytochem Rev 16:1227–1252
11. Shlaes DM (2010) The perfect storm. Antibiotics. Springer, Dordrecht, pp 1–7
12. Brown GD, Denning DW, Gow NA, Levitz SM, Netea MG, White TC (2012) Hidden killers: human fungal infections. Sci Trans Med 4:165rv13-rv13
13. Nidiry ESJ, Ganeshan G, Lokesha AN (2011) Antifungal activity and isomerization of octadecyl p-coumarates from Ipomoea carnea subsp. fistulosa. Nat Prod Commun 6:1889–1892
14. De Morais C, Scopel M, Pedrazza G, da Silva F, Dalla Lana D, Tonello M et al (2017) Antidermatophyte activity of Leguminosae plants from Southern Brazil with emphasis on Mimosa pigra (Leguminosae). J Mycol Med 27:530–538
15. Isa AI, Awouafack MD, Dzoyem JP, Aliyu M, Magaji RA, Ayo JO et al (2014) Some Strychnos spinosa (Loganiaceae) leaf extracts and fractions have good antimicrobial activities and low cytotoxicities. BMC Complement Altern Med 14:456

Chapter 15

Screening and Assessment of Pisatin Demethylase Activity (PDA)

Catherine C. Wasmann and Jeffrey J. Coleman

Abstract

Plants produce low molecular weight compounds with antimicrobial activity in response to microbial attack termed phytoalexins. The first phytoalexin identified was (+) pisatin from pea, and several fungi are able to detoxify pisatin to a less inhibitory compound, including *F. oxysporum* f. sp. *pisi*. This detoxification is catalyzed by demethylation of the compound (termed pisatin demethylase activity, or PDA) by the cytochrome P450, Pda. Here we detail two procedures to assess PDA using radiolabeled [^{14}C]pisatin as a substrate and monitoring activity using a scintillation counter.

Key words Host-specific virulence factor, Pea, Phytoalexin, *F. oxysporum* f. sp. *pisi*

1 Introduction

Once infected with a pathogen, plants have many defense responses including the synthesis of low molecular weight antimicrobial compounds termed phytoalexins. The first phytoalexin identified was (+) pisatin from pea [1], and soon after its initial discovery a correlation was found between the tolerance of a pathogen to pisatin and the ability to cause disease on pea [2]. Since then several fungi have shown to have the ability to detoxify pisatin to the less inhibitory compound (+)6a-hydroxymaackiain (Fig. 1) [3, 4], including *F. oxysporum* f. sp. *pisi* [5]. The cytochrome P450 pisatin demethylase (Pda) is responsible for this catalysis, and has been best studied in isolates of *F. vanettenii* (formerly, *Nectria haematococca* MPVI) [6, 7]. In *F. vanettenii*, *PDA1* is a virulence factor for garden pea, as disruption of the gene results in decreased virulence [8–10].

Homologs of *PDA* have been identified in several *F. oxysporum* isolates [11], although it was only functional within *F. oxysporum* f. sp. *pisi* isolates [5]. Here we detail two procedures to assess PDA using radiolabeled [^{14}C]pisatin as a substrate (Fig. 1) and

Jeffrey Coleman (ed.), *Fusarium wilt: Methods and Protocols*,
Methods in Molecular Biology, vol. 2391, https://doi.org/10.1007/978-1-0716-1795-3_15,

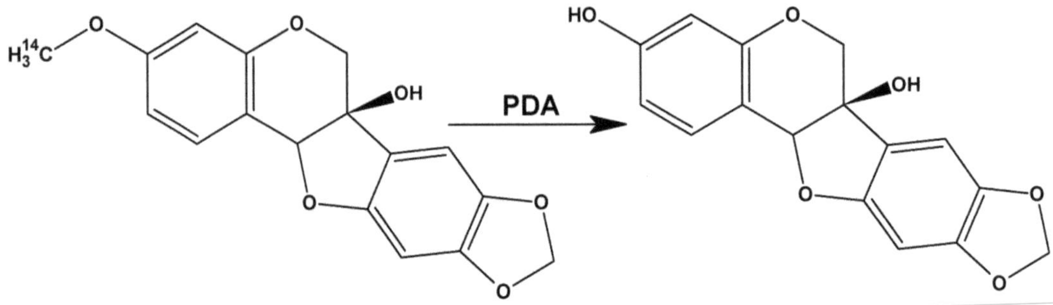

Fig. 1 Enzymatic detoxification of (+)pisatin mediated by pisatin demethylase (Pda). The [^{14}C] labeled carbon is indicated, and its loss is monitored in these assays

monitoring activity using a scintillation counter. The first assay enables multiple isolates/strains to be assessed for pisatin demethylase activity, while the second assay provides a method to monitor the enzymatic activity over time.

2 Materials

2.1 Biological Samples

1. *F. oxysporum* f. sp. *pisi* isolate(s) or other fungi that are to be assayed (*see* **Note 1**).

2.2 Screening of Pisatin Demethylase Activity (Scintillation Vial Assay)

1. M-2 medium: Mix 10 g dextrose, 5 g peptone, 0.5 g yeast extract, 0.5 g KH_2PO_4, 0.5 g K_2HPO_4, 0.5 g $MgSO_4{\cdot}7\ H_2O$, and 20 g agar. Autoclave, and allow to cool.
2. (+) pisatin (*see* **Note 2**).
3. [^{14}C] labeled pisatin.
4. Dimethylsulfoxide (DMSO).
5. Sterile scintillation vials with lids.
6. Sterile wooden applicators.
7. Plastic bag.
8. Scintillation fluid: 0.55% 2,5-diphenyloxazole (PPO) in toluene.
9. Scintillation counter.

2.3 Assessment of Pisatin Demethylase Activity over Time

1. Sterile 50 mL Erlenmeyer flasks.
2. GY medium [12]: Mix 50 g glucose, 8 g casein hydrolysate, 0.5 g yeast extract, 20 mg $MgSO_4{\cdot}7\ H_2O$, 2 mg $CaCl_2{\cdot}2\ H_2O$, 2 mg $FeSO_4{\cdot}7\ H_2O$, 75 μg $MnSO_4{\cdot}4\ H_2O$, 75 μg $Na_2MoO_2{\cdot}2\ H_2O$, in 1000 mL 0.05 potassium phosphate buffer (pH 7.5). Autoclave.

3. 0.05 M potassium phosphate buffer (pH 6.5): Dissolve 2.94 g K_2HPO_4 and 4.51 g KH_2PO_4 in 800 mL ddH_2O. Bring final volume to 1 L and sterilize by autoclave.
4. Sterile funnels with Nitex (30 μm pore size).
5. (+) pisatin.
6. [^{14}C] labeled pisatin.
7. DMSO.
8. Shaking water bath.
9. Sterile scintillation vials with lids.
10. Scintillation fluid (same as above).
11. Scintillation counter.

3 Methods

3.1 Screening of Pisatin Demethylase Activity (Scintillation Vial Assay)

1. Prepare assay scintillation vials by mixing M-2 medium, (+) pisatin, [^{14}C]pisatin, and DMSO on a warming plate to make M-2 medium with 0.1 mM pisatin at 5000 cpm/mL (31 μg pisatin/5000 cpm/10 μL DMSO/mL medium; *see* **Note 3**).
2. Pipet 250 μL of the medium containing the [^{14}C]pisatin to the bottom of a sterile scintillation vial, and cover to allow to solidify (*see* **Note 4**).
3. Once solidified, inoculate the fungi to be assayed to each vial. Using a sterile wooden applicator, collect conidia by touching the end of the applicator to a sporulating culture. Gently touch the inoculated end of the applicator to the center of the M-2 medium containing [^{14}C]pisatin. It is recommended to inoculate a *F. oxysporum* isolate that is known to have PDA to serve as a positive control, as well as an isolate that does not have PDA to serve as a negative control. An uninoculated vial can serve as a negative control if no alternative is available. Cap the vials loosely and label with the inoculate information (strain, mutant, etc.).
4. Place all inoculated vials into a plastic bag containing moist paper towels to prevent drying.
5. Place bag containing inoculated vials at 25 °C for 5–7 days for growth (*see* **Note 5**).
6. Add 4.5 mL of scintillation fluid to each of the vials, and check to make sure growth occurred. Let the vials stand 1 h to overnight.
7. Measure the amount of [^{14}C] present in the vials using a scintillation counter.

3.2 Assessment of Pisatin Demethylase Activity over Time

1. Inoculate 50 mL of GY medium in a 250 mL Erlenmeyer flask with conidia of the *F. oxysporum* isolate with PDA to be assayed. Allow to grow at room temperature on a rotary shaker for 40 h.
2. Collect mycelia by filtering through Nitex (30 μm), and wash with 0.05 potassium phosphate buffer (*see* **Note 6**).
3. The washed mycelium is suspended in 20 mL of the 0.05 potassium phosphate buffer at a concentration of 30 mg/mL (*see* **Note 7**).
4. [^{14}C] labeled pisatin (300 dpm/μg) was added to a final concentration of 0.1 mM pisatin in 0.5% DMSO.
5. The flask is placed in a 25 °C shaking water bath at 100 rpm for the duration of the assay.
6. Aliquots (500 μL) are removed every 15–30 min over the duration of the experiment and added to scintillation vials (*see* **Note 8**).
7. To the scintillation vial, add 4.5 mL of scintillation fluid and briefly vortex.
8. Measure the amount of [^{14}C] present in the vials using a scintillation counter.
9. Plot the percent of pisatin remaining (or pisatin concentration) versus time (h).

4 Notes

1. As pisatin is produced by pea plants, most experiments would likely involve assaying *F. oxysporum* f. sp. *pisi* isolates. Isolates are available from the Fungal Genetics Stock Center and the ARS culture collection.
2. (+)Pisatin is not commercially available and needs to be extracted and purified in the laboratory or be obtained from collaborators. While treatment with pathogens elicits the synthesis of pisatin, treatment of pea germlings with 5–10 mM $CuCl_2$ for 1 h induces production of the compound.
3. A higher concentration of pisatin can be used, but the concentration should be non-inhibitory to the isolates. Additionally a higher amount of [^{14}C]pisatin can be used, but it is not recommended to go above 20,000 cpm/mL of medium. Additionally, as little as 1000 cpm/mL of [^{14}C]pisatin can be used in order to conserve the labeled compound.
4. It is important to ensure all of the medium is on the bottom of the vial and not adhering to the sides. Medium containing [^{14}C]pisatin on the sides of the vial will not come in contact with the fungus and therefore will not be demethylated. This

could result in an isolate with PDA activity to appear negative in this assay.

5. Fungal growth can be monitored in the vial by observing through the bottom of the vial when held to a light source. Frequently the purple pigment known as bikaverin that is produced by *F. oxysporum* isolates [13] is viable through the vial and can serve as a proxy for fungal growth. It should be noted that bikaverin production is not always apparent despite fungal growth.
6. It is very important that the mycelia collected on the Nitex filter is not over dried, as the mycelia should be very moist enabling it to be easily suspended in the potassium phosphate buffer. To correct for the inaccuracy this adds to the assay, dry weights should be conducted on the suspended mycelial culture.
7. Depending on the fungal isolate being assayed, glucose could repress PDA. For instance, glucose represses PDA in *F. vanettenii* but has no effect on PDA in *Ascochyta pisi*. There is delayed PDA in *F. oxysporum*, and therefore assays with 1% glucose and without glucose could be conducted to determine the effect of the sugar.
8. A volume of as little as 250 μL can be removed and placed in scintillation vials for analysis. After 8 h of co-incubation of *F. oxysporum* with the [^{14}C] labeled pisatin, there is usually no radioactive pisatin remaining, and it has all been converted to 6a-hydroxymaackiain. Some assays are finished in as little as 3 h, so it is important to continually monitor the experiment.

References

1. Cruickshank IAM, Perrin DR (1960) Isolation of a phytoalexin from *Pisum sativum* L. Nature 187:799–800
2. Cruickshank I (1962) Studies on phytoalexins IV. The antimicrobial spectrum of pisatin. Aust J Biol Sci 15:147–159
3. Delserone LM, McCluskey K, Matthews DE, Vanetten HD (1999) Pisatin demethylation by fungal pathogens and nonpathogens of pea: association with pisatin tolerance and virulence. Physiol Mol Plant Pathol 55:317–326
4. George HL, VanEtten HD (2001) Characterization of pisatin-inducible cytochrome P450s in fungal pathogens of pea that detoxify the pea phytoalexin pisatin. Fungal Genet Biol 33:37–48
5. Coleman JJ, Wasmann CC, Usami T, White GJ, Temporini ED, McCluskey K, VanEtten HD (2011) Characterization of the gene encoding pisatin demethylase (*FoPDA1*) in *Fusarium oxysporum*. Mol Plant-Microbe Interact 24:1482–1491
6. Vanetten H, Temporini E, Wasmann C (2001) Phytoalexin (and phytoanticipin) tolerance as a virulence trait: why is it not required by all pathogens? Physiol Mol Plant Pathol 59:83–93
7. Geiser DM, Al-Hatmi A, Aoki T, Arie T, Balmas V, Barnes I, Bergstrom GC, Bhattacharyya MKK, Blomquist CL, Bowden R, Brankovics B, Brown DW, Burgess LW, Bushley K, Busman M, Cano-Lira JF, Carrillo JD, Chang HX, Chen CY, Chen W, Chilvers MI, Chulze SN, Coleman JJ, Cuomo CA, de Beer ZW, de Hoog GS, Del Castillo-Múnera J, Del Ponte E, Diéguez-Uribeondo J, Di Pietro A, Edel-Hermann V, Elmer WH, Epstein L, Eskalen A, Esposto MC, Everts KL, Fernández-Pavía SP, da Silva GF, Foroud NA, Fourie G, Frandsen RJN, Freeman S, Freitag M, Frenkel O, Fuller KK, Gagkaeva T,

Gardiner DM, Glenn AE, Gold S, Gordon T, Gregory NF, Gryzenhout M, Guarro J, Gugino B, Gutiérrez S, Hammond-Kosack K, Harris LJ, Homa M, Hong CF, Hornok L, Huang JW, Ilkit M, Jacobs A, Jacobs K, Jiang C, Jimenez-Gasco MDM, Kang S, Kasson MT, Kazan K, Kennell JC, Kim H, Kistler HC, Kuldau GA, Kulik T, Kurzai O, Laraba I, Laurence MH, Lee TY, Lee YW, Lee YH, Leslie JF, Liew ECY, Lofton LW, Logrieco A, Sánchez López-Berges M, Luque AG, Lysøe E, Ma LJ, Marra RE, Martin FN, May SR, McCormick S, McGee CT, Meis JF, Migheli Q, Mohamed Nor NMI, Monod M, Moretti A, Mostert D, Mulé G, Munaut F, Munkvold GP, Nicholson P, Nucci M, O'Donnell K, Pasquali M, Pfenning LH, Prigitano A, Proctor R, Ranque S, Rehner S, Rep M, Rodríguez-Alvarado G, Rose LJ, Roth MG, Ruiz-Roldán C, Saleh AA, Salleh B, Sang H, Scandiani M, Scauflaire J, Schmale D 3rd, Short DP, Šišić A, Smith J, Smyth CW, Son H, Spahr E, Stajich JE, Steenkamp E, Steinberg C, Subramaniam R, Suga H, Summerell BA, Susca A, Swett CL, Toomajian C, Torres-Cruz TJ, Tortorano AM, Urban M, Vaillancourt LJ, Vallad GE, van der Lee T, Vanderpool D, van Diepeningen AD, Vaughan M, Venter E, Vermeulen M, Verweij PE, Viljoen A, Waalwijk C, Wallace EC, Walther G, Wang J, Ward T, Wickes B, Wiederhold NP, Wingfield MJ, Wood AKM, Xu JR, Yang XB, Yli-Matilla T, Yun SH, Zakaria L, Zhang H, Zhang N, Zhang S, Zhang X (2020) Phylogenomic analysis of a 55.1 kb 19-gene dataset resolves a monophyletic *Fusarium* that includes the *Fusarium solani* Species Complex. Phytopathology 111: 1064–1079

8. Wasmann CC, VanEtten HD (1996) Transformation-mediated chromosome loss and disruption of a gene for pisatin demethylase decrease the virulence of *Nectria haematococca* on pea. Mol Plant-Microbe Interact 9:793–803
9. Coleman JJ, White GJ, Rodriguez-Carres M, VanEtten HD (2011) An ABC transporter and a cytochrome P450 of *Nectria haematococca* MPVI are virulence factors on pea and are the major tolerance mechanisms to the phytoalexin pisatin. Mol Plant-Microbe Interact 24:368–376
10. Ciuffetti LM, VanEtten HD (1996) Virulence of a pisatin demethylase-deficient *Nectria haematococca* MPVI isolate is increased by transformation with a pisatin demethylase gene. Mol Plant-Microbe Interact 9:787–792
11. Temporini ED, VanEtten HD (2004) An analysis of the phylogenetic distribution of the pea pathogenicity genes of *Nectria haematococca* MPVI supports the hypothesis of their origin by horizontal transfer and uncovers a potentially new pathogen of garden pea: *Neocosmospora boniensis*. Curr Genet 46:29–36
12. Van Etten HD, Barz W (1981) Expression of pisatin demethylating ability in *Nectria haematococca*. Arch Microbiol 129:56–60
13. Wang Q, Cobine PA, Coleman JJ (2018) Efficient genome editing in *Fusarium oxysporum* based on CRISPR/Cas9 ribonucleoprotein complexes. Fungal Genet Biol 117:21–29

Chapter 16

Detection and Genotyping of Fov4 (Race 4, VCG0114), the *Fusarium* Wilt Pathogen of Cotton

Jinggao Liu and Tanya A. Wagner

Abstract

Fusarium wilt, caused by *Fusarium oxysporum* f. sp. *vasinfectum* (Fov), is an important disease of cotton. More than 14 different genotypes as determined by VCG and sequence analyses are known to occur in the United States. Fov4 (race 4, VCG0114), originally found in India, was first detected in the United States in 2001 in California and recently in 2017 and 2019 in Texas and New Mexico, respectively. Four sub-genotypes of Fov4 have been identified, with Fov4 N, T, and MiT genotypes occurring in California, and Fov4 T and MT genotypes occurring in Texas. Unlike other genotypes of Fov in the United States, Fov4 does not require the presence of root-knot nematodes (*Meloidogyne incognita*) to cause severe wilt in cotton and is a major concern to US cotton growers. Fov4 can be spread through a variety of mechanisms including through infected seed. Once a field is infested, the fungus becomes endemic since there are no economically viable means to eradicate the pathogen from infested fields. Therefore, a rapid and accurate detection method is essential for early identification of infested fields and seed lots to prevent further spread of Fov4. This chapter describes multiplex and singleplex PCR diagnostics for detection of Fov4, and for detection and genotyping N, T, MiT, and MT genotypes of Fov4 from wilted cotton plants.

Key words *Fusarium* wilt, *Fusarium oxysporum* f. sp. *vasinfectum*, Cotton, Diagnostics, Detection, Fov4, Race 4, VCG0114, Multiplex, PCR

1 Introduction

Fusarium wilt of cotton (*Gossypium* spp.), caused by the soilborne fungus *Fusarium oxysporum* f. sp. *vasinfectum* (Fov), is an economically important disease occurring in most cotton-growing regions of the world. Historically, variations in pathogenicity to diverse plant species as well as *Gossypium* species and cultivars were used to group Fov isolates into races 1 to 8 [1–4]. Genotypic characterization by DNA sequence and vegetative compatibility analyses in conjunction with pathogenicity assays revealed the polyphyletic nature of the known Fov races and redundancies between race 3 and race 5, and between race 4 and race 7 [5–10]. Furthermore,

Jeffrey Coleman (ed.), *Fusarium wilt: Methods and Protocols*,
Methods in Molecular Biology, vol. 2391, https://doi.org/10.1007/978-1-0716-1795-3_16,

these race designations have been shown to be invalid as well as impractical [9, 11–14]. Subsequently identified genotypes of Fov were no longer assigned to new races [9, 11, 12]. In the present study, race terminology is retained to refer the genotypes of the historical isolates.

Vegetative compatibility refers to the ability of hyphae of two isolates to anastomose and form a stable heterokaryon. This process in filamentous fungi is controlled by a dozen or more vegetative incompatibility (*vic*) loci, and only isolates with identical alleles can complement each other [15]. Thus, vegetative compatibility analyses allow grouping of Fov isolates into genetically distinctive vegetative compatibility groups (VCG). More than 21 VCGs and two distinct pathotypes have been identified [5, 9, 12, 16]. Isolates of VCG0113 (races 3 and 5), 0114 (Fov4, races 4 and 7), and 01111 (Australian biotype) are capable of causing disease in the absence of nematodes and are often associated with heavy clay or loam soils [12, 13, 17]. They cause severe wilt and extensive root damage in a soil-infestation assay but fail to cause disease when inoculated into the cotton vasculature in a stem-puncture assay; thus they are classified as a nematode-independent, vascular-incompetent, or root-rot pathotype [9, 18]. Control of these nematode-independent pathotype isolates requires deployment of host resistance to the fungus itself. Isolates of other VCGs, including the named races 1 (VCG0111), 2 (VCG0112), 6 (VCG0116), and 8 (VCG01119), are usually found in sandy acidic soils and require the presence of nematodes such as the root-knot nematode *Meloidogyne incognita* (Chitwood) to cause appreciable disease [19–21]. These isolates cause wilt symptoms in the stem-puncture assay but not in the soil-infestation assay unless nematodes are included in the assay. Consequently, these isolates are classified as a vascular-competent or nematode-dependent pathotype [9]. Control of these nematode-dependent pathotype isolates has been mainly achieved by breeding for resistance to the nematodes [22].

More than 13 VCGs belonging to the nematode-dependent pathotype are known to occur in the United States, with VCG0112 (race 2), 01117 (not race 1, but recognized as race 1 based on DNA similarity), and 01119 (race 8) as the most widely occurring genotypes [5, 7, 9, 23]. Fov4, the only known genotype of Fov belonging to the nematode-independent pathotype in the United States, was detected in the San Joaquin Valley of California in 2001 [13]. The California Fov4 isolates vegetatively complemented the VCG0114 reference isolates from India (race 4) and China (race 7) [5]. Isolates with *EF-1a* sequences identical to those of Fov4 were also reported from Mid-South states [7, 24]. These isolates were designated as "Race 4-like" or the MDS-12 genotype. Some of these isolates had DNA sequences of the *EF-1a*, *PHO*, and *BT* genes identical to those of Fov4 isolates [10]. However, they failed to vegetatively complement Fov4 isolates or isolates of any known Fov VCGs, and were found to be non-pathogenic to cotton [9, 10, 24].

Fov4 is of great concern to growers because it is highly virulent on many cotton cultivars, especially certain Pima cultivars, and there are few Fov4-resistant commercial cultivars [13, 25]. Fov4 has spread throughout the San Joaquin Valley, and was first detected outside California in 2017 in Texas, the largest cotton-growing state in the United States [26]. Subsequently, Fov4 was confirmed in New Mexico in 2019 [27]. The pathogen is known to spread through many possible mechanisms including transmission by seed, insect, nematode, and through movement of infested soil and plant debris [14, 28–31]. Once established, the pathogen persists in the field for decades, even in the absence of the cotton host [32], and there are no effective means to eradicate the pathogen from infested fields. A reliable rapid detection method is essential for early identification of infested fields and seed lots to prevent further spread of Fov4.

A PCR-based detection method using Fov4-specific primers based on unique AFLP polymorphisms [33] and a commercial detection kit using the same primer but modified for the isothermal recombinase polymerase amplification platform have been developed [34]. However, both methods give false positive reactions for the VCG0113 (race 3) isolates [35]. Recently, four genotypes (N, T, MT, and MiT) were identified among Fov4 isolates based on the various insertions of the Class II transposable elements *Tfo1*, *MULE*, and *MITE* into the *PHO* locus (Fig. 1, left panel), and a multiplex PCR method (Fig. 1, right panel) to specifically detect each of these genotypes was developed [5, 10]. The four genotypes can also be detected in three singleplex PCR reactions targeting N-, T/MiT-, and MT-specific regions. Furthermore, an Fov4-specific PCR detection method was developed based on a *Tfo1* insertion event at the *NAT* gene locus (Fig. 1, left panel) [10]. This method provides a rapid alternative approach to vegetative compatibility analysis for identifying Fov4, which is labor-intensive and time-consuming. The specificity of the N genotype determination is contingent upon a positive result with this Fov4-specific test targeting the *NAT* locus. Using these approaches, we identified Fov4 T and MT genotypes in Texas, N and MT genotypes in China, and N, T, and MiT genotypes in California [10]. Subsequently others reported Fov4 N and MT genotypes in New Mexico, and N and T genotypes in California utilizing these approaches [36, 37].

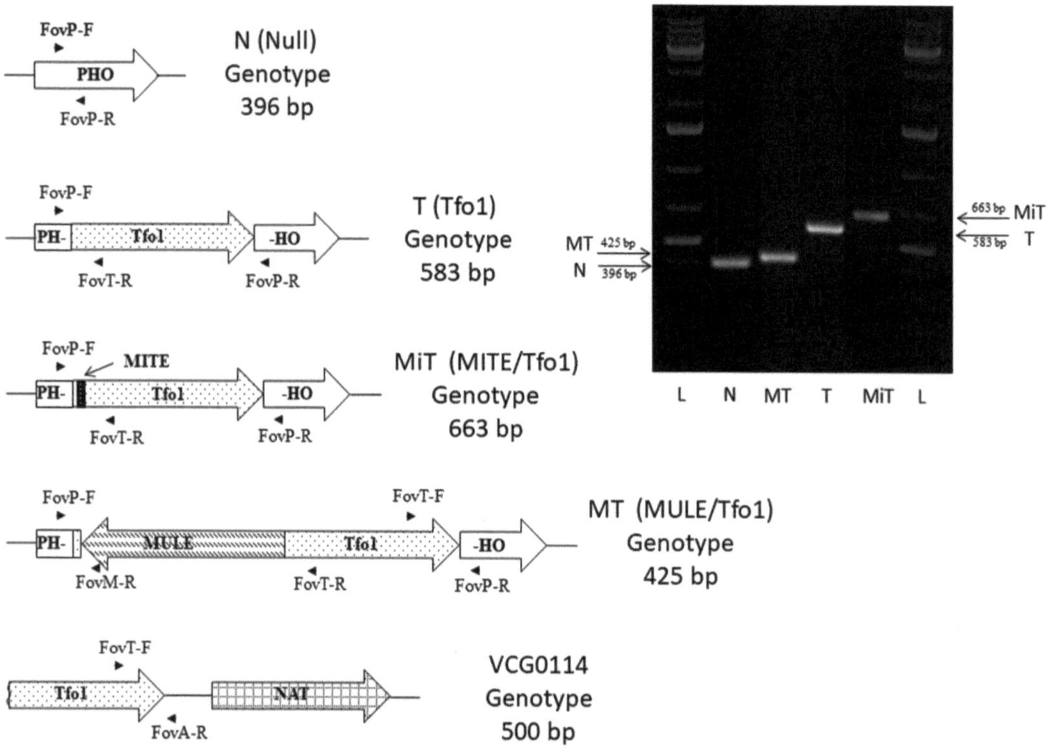

Fig. 1 Genotype determinations of *F. oxysporum* f. sp. *vasinfectum* VCG0114 (Fov4, race 4) isolates and their specific detection. Left panel: four genotypes (N, T, MT, and MiT) observed among VCG0114 isolates based on the various transposable element insertion events at the *PHO* locus and their specific detections. The genotype of Fov4 is determined and specifically detected based on the unique *Tfo1* insertion event at the N-acetyl transferase gene (*NAT*) locus using the primer pair FovT-F and FovA-R producing an amplicon size of 500 bp. The N (null) genotype determination is not specific unless the Fov4 genotype determination is positive at the *NAT* locus. Target location of PCR primers is indicated on the diagrams. Right panel: an example of genotype determination by a multiplex PCR with the primer mixture of FovP-F, FovP-R, FovT-R, and FovM-R. Lane L, 1 kb ladder DNA plus size marker; lane N (N genotype), Indian isolate ATCC16613; lane MT (*MULE/Tfo1* genotype), Texas isolate TX11; lane T (*Tfo1* genotype), Texas isolate TX39; and lane MiT (*MITE/Tfo1* genotype), California isolate MD308 (reproduced from ref. [10], which is in the public domain)

2 Materials

2.1 Culture Media and Isolation

1. Komada's medium [38] (*see* **Note 1**): Dissolve 1 g K_2HPO_4, 0.5 g KCl, 0.5 g $MgSO_4{\cdot}7\ H_2O$, 0.01 g Fe-Na-EDTA, 20 g D-galactose, 2 g L-asparagine, 20 g agar, 1 g Terraclor (75% pentachloronitrobenzene (PCNB)) in 800 mL H_2O. Adjust pH to 3.6–4.0 with 10% phosphoric acid, and bring to 1 L with H_2O. Autoclave at 121 °C for 15 min, and cool the autoclaved mixture to 50 °C. Add 10 mL of filter sterilized oxgall-$Na_2B_4O_7{\cdot}\ H_2O$ solution (0.5 g and 1 g, respectively, dissolved in 10 mL of water) and 6 mL of filter sterilized streptomycin sulfate solution (0.3 g dissolved in 6 mL of

water). Dispense into Petri dishes with occasional gentle shaking of the flask (*see* **Note 2**), and cool to room temperature (*see* **Note 3**).

2. Potato dextrose agar amended with 100 ppm chloramphenicol and 50 ppm tetracycline (PDAA): Mix 39 g potato dextrose agar in 1 L distilled water. Autoclave at 121 °C for 15 min, and cool the autoclaved medium to 50 °C. Add 10 mL of the chloramphenicol-tetracycline stock solution and mix well. Dispense into Petri dishes and cool to room temperature (*see* **Note 3**).
3. Chloramphenicol-tetracycline stock solution: Dissolve 0.5 g chloramphenicol and 1 g tetracycline in 100 mL of 70% ethanol. Filter sterilize and store at −20 °C.
4. Yeast extract-peptone-dextrose (YEPD or YPD) broth: Suspend 10 g yeast extract, 20 g peptone, and 20 g dextrose in 1 L of distilled water. Autoclave at 121 °C for 15 min, and allow to cool to room temperature. The medium can be stored at 4 °C for several weeks.
5. Brown paper bags.
6. Sterile scalpel.
7. Forceps.
8. 70% ethanol.
9. Sterile water.
10. Sterile paper towels.
11. Dissecting microscope.
12. Cell spreader.
13. Cryotubes.
14. 20% sterile glycerol.

2.2 DNA Extraction

1. Lysis Buffer: 100 mM LiCl, 10 mM EDTA, 10 mM Tris (pH 8.0), 0.5% sodium dodecylsulfate. Immediately before use, adjust solution to 0.1% β-mercaptoethanol (v/v).
2. Lysis Buffer C: 2 M NaCl, 0.2 M Tris (pH 8.0), 50 mM EDTA, 55 mM hexadecyl triammonium bromide (CTAB). This buffer can be stored for 1–2 years. Immediately before use, remove an aliquot, add β-mercaptoethanol (v/v) to 0.3%, and prewarm to 65 °C.
3. 5% Sarkosyl (N-Lauroylsarcosine, sodium salt) in water.
4. Phenol, pH 8.0.
5. Chloroform-isoamyl alcohol (v/v, 24:1).
6. 1× Low TE Buffer.
7. 50 mL sterile culture tubes.

8. Sterile toothpicks.
9. Sterile disposable polypropylene pestles.
10. Microcentrifuge tubes.
11. Vortex.
12. Dry heating block or water bath.
13. Microcentrifuge.
14. 3 M sodium acetate, pH 5.2 (NaOAc).
15. Ethanol.
16. Nanodrop.

2.3 PCR Detection of Fov4-Specific Polymorphisms

1. 5× Green GoTaq Flexi Buffer
2. 25 mM $MgCl_2$.
3. 10 mM dNTPs.
4. GoTaq DNA polymerase.
5. Primers for specific Fov4 polymorphisms (Table 1).
6. Genomic DNA of each Fov4 genotype to use as a positive control.

Table 1
Primers used for Fov4 detection and genotyping by PCR [10]

Name	Sequence (5′-3′)
Detection of ***Fov4*** *(VCG0114, race 4) genotype (500 bp)*	
FovT-F	GTCAGGCTCGCGAGGAAATA
FovA-R	TCAACAGACCCTGCACTACG
Detection of ***Fov4*** *N genotype (396 bp)*	
FovP-F	GGCCGATATTGTCGGTCGTA
FovP-R	CTCCAGTGCAGTGCTTGGTA
Detection of ***Fov4*** *T (583 bp) or* ***Fov4*** ***MiT*** *genotype (663 bp)*	
FovP-F	GGCCGATATTGTCGGTCGTA
FovT-R	ATCTGTCTTTCGTCGGCAAT
Detection of ***Fov4*** ***MT*** *genotype (425 bp)*	
FovP-F	GGCCGATATTGTCGGTCGTA
FovM-R	CCGCCATATCCACTGAACA
Multiplex detection of ***Fov4*** *N, T, MiT or MT genotype*	
FovP-F	GGCCGATATTGTCGGTCGTA
FovP-R	CTCCAGTGCAGTGCTTGGTA
FovT-R	ATCTGTCTTTCGTCGGCAAT
FovM-R	CCGCCATATCCACTGAACA

2.4 Gel Electrophoresis

1. 50× TAE Buffer (Tris-Acetate-EDTA).
2. 10 mg/mL ethidium bromide solution.
3. 1 kB DNA ladder.
4. Agarose.
5. Gel casting/running system with power supply.
6. Gel documentation system.

3 Methods

3.1 Sample Collection

1. Scout suspected cotton fields between the first true leaf and first bloom stages (*see* **Note 4**) for empty spots where seedlings or plants are killed or stunted. Identify plants displaying wilt symptoms: epinasty, chlorosis and necrosis of leaves, defoliation, stunting, and death.
2. Pull the wilted and symptomatic (*see* **Note 5**) plants from the soil, and remove sections 2 in. above and below soil line (*see* **Note 6**). Place the middle sections of the plants in brown paper bags, and keep the bags in a ventilated, cool, and dry area before being brought to the lab as soon as possible (*see* **Note 7**).

3.2 Selective Plating

1. Collected samples need to be processed as soon as possible. Surface sterilize the tap root-stem samples by immersing in 70% ethanol for 3 min and rinsing with sterile water.
2. Peel the bark off and trim the ends with a sterile scalpel and forceps, immerse again in 70% ethanol for 1 min, rinse with sterile water, blot dry with a sterile paper towel, and cut into halves.
3. Cut both the tap root and stem portions lengthwise. Press these half pieces slightly into the agar surface of Komada's medium with the cut face embedded to the medium (less than 1 mm). Incubate the plates at 25 °C for 5–7 days (*see* **Note 8**).
4. Observe the plates under a dissecting microscope for the presence of *F. oxysporum*-like colonies and sporulation according to Nelson et al. [39]. Fov4 colonies often display floccose areal mycelium and range in color from cream white to light purple. There is a wide variation in the colony morphology of individual isolates.
5. Purify tentative *F. oxysporum* colonies by dilution plating on PDAA plates by scraping a pea-sized culture from the relatively densely growing area just behind the edge of the colony, and make a ten-fold dilution series in sterile water. Spread 50 μL of each dilution on a PDAA plate. After 5–7 days, pick a well-

separated Fov colony and use it as a single-spore culture. Long-term storage of cultures can be made with conidia suspended in 20% sterile glycerol and can be kept at −80 °C for several years.

3.3 DNA Extraction

This method is adapted from ref. [40]. Multiple samples (12 to 16) can be batch processed.

3.3.1 DNA Extraction from Fungal Cultures

1. Inoculate 10 mL of YEPD broth with a single spore culture using a sterile toothpick, and incubate at 28 °C for 4–5 days without shaking.
2. Collect the mycelium with a sterile toothpick and blot out the medium with paper towels.
3. Grind mycelial tissue (80 mg) to a fine power with a disposable polypropylene pestle in a 1.5 mL microcentrifuge tube chilled in liquid nitrogen (*see* **Note 9**).
4. Add 800 μL of Lysis Buffer and vortex.
5. Incubate at 65 °C for 10 min. Mix 2 or 3 times during incubation by inverting the tube.
6. Add 800 μL of phenol (pH 8), vortex 1 min, and centrifuge at high speed (microcentrifuge) for 5 min.
7. Transfer 600 μL of the top aqueous layer to a new tube, add 200 μL of Lysis Buffer, 250 μL of phenol (pH 8), 250 μL of chloroform-isoamyl alcohol (24:1), vortex 1 min, and centrifuge at high speed for 5 min.
8. Transfer 600 μL of the top aqueous layer to a new tube, add 500 μL of the chloroform-isoamyl alcohol, vortex 1 min, and centrifuge at high speed for 5 min.
9. Transfer 400 μL of the top aqueous layer to a new tube, add 40 μL of 3 M NaOAc and 800 μL of 100% ethanol, mix by inversion, and centrifuge at top speed for 5 min.
10. Decant the supernatant without dislodging the pellet, add 1 mL of 70% ethanol, vortex 1 min (*see* **Note 10**), and centrifuge at high speed for 5 min.
11. Decant the supernatant without dislodging the pellet, invert the open tubes on several layers of paper towels, and air dry the samples for 30 min (*see* **Note 11**).
12. Redissolve the DNA pellet in 50 μL 1× Low TE Buffer.
13. Determine DNA concentration using a Nanodrop, and check the DNA integrity by gel electrophoresis.
14. Dilute the DNA with molecular grade water to make a 20 ng/μ L working solution.

3.3.2 DNA Extraction from Cotton

This method is adapted from ref. [41]. Multiple samples (12 to 16) can be batch processed.

DNA can be directly extracted from freshly collected tap root-stem samples from wilted cotton plants to determine and detect Fov4 genotypes. However, this method is less sensitive than detection using DNA extracted from fungal isolates recovered by selective plating [5]. Tissue samples from young seedlings with obvious vascular staining are more successful than samples without vascular staining.

1. Cut the tap-root or stem tissue of the collected sample into thin slices, and place the cut tissue (80 mg) into a 1.5 mL microcentrifuge tube prechilled in liquid nitrogen.
2. Grind the tissue to a fine powder with a disposable polypropylene pestle in liquid nitrogen.
3. Add 600 μL of prewarmed Lysis Buffer C, vortex 1 min, and place on a rack until the rest of the samples are processed.
4. Add 60 μL of 5% sarkosyl to each tube and mix by inversion.
5. Incubate at 65 °C for 15 min, and mix by inverting tubes 2 or 3 times during incubation.
6. Add 600 μL of chloroform-isoamyl alcohol (24:1) to each tube, and mix by vortex and vigorous inversion.
7. Centrifuge at top speed for 5 min.
8. From each tube, transfer 550 μL of the supernatant to a new tube, and avoid the cloudy and thick material debris between the two phases.
9. Add 550 μL of isopropanol to each tube, mix by inverting the tubes 10 times, and centrifuge at top speed for 5 min.
10. Decant the supernatants without dislodging the DNA pellets from each tube.
11. Add 700 μL of 70% ethanol to each tube, vortex 1 min (*see* **Note 10**), and centrifuge at top speed for 5 min.
12. Decant the supernatants without dislodging the pellets, invert the open tubes on several layers of paper towels, and air dry the samples for 30 min (*see* **Note 11**).
13. Redissolve the DNA pellet in 50 μL of 1× Low TE Buffer.
14. Determine the DNA concentration using a Nanodrop, and check the DNA integrity by gel electrophoresis.
15. Dilute the DNA with molecular grade water to make a 200 ng/μL working solution.

3.4 Detection and Genotype Determination of Fov4

For detection and genotyping Fov4 N, T, MT, and MiT genotypes, use either template DNA working solutions prepared from isolated fungal cultures (20 ng/mL) or directly from infected plant tissues (200 ng/mL). Primers FovP-F, FovP-R, FovT-R, and FovM-R (Table 1) are used for this multiplex PCR.

3.4.1 Detection and Genotype Determination of Fov4 by Multiplex PCR

1. Assemble the PCR reactions on ice. Prepare a master mix excluding the template DNA as listed in Table 2. Prepare enough master mix for the number of samples to be tested (include an extra 10% of master mix to account for pipetting error). Dispense 19 μL of the master mix into each PCR tube.
2. Add 1 μL of the template DNA (working solution) to each tube.
3. A negative control with 1 μL of ddH_2O and positive controls with 1 μL of DNA (20 ng/μL) from a known Fov4 isolate for each of the four Fov4 genotypes should be included.
4. Run the multiplex PCR on a thermal cycler with the following program: 94 °C for 2 min; followed by 35 cycles of 94 °C for 30 s, 58 °C for 30 s, 72 °C for 40 s; with a final extension of 72 °C for 5 min and a 4 °C hold.
5. Load 10 μL of each PCR reaction and several lanes of a 1 kb DNA Ladder (if Green GoTaq Flexi Buffer or similar is not used, add 2 μL of 6× loading dye to each product; place 1 kb DNA Ladders in the first lane and the middle lane bordering the positive controls) to a 1.7% agarose gel in 1× TAE buffer containing ethidium bromide (0.2 μg/mL final concentration; *see* **Note 12**), and run at 6 V/cm until the dye front reaches the bottom of the gel (at least 8 cm, 60 min).

Table 2
Reagents for PCR (20 μL reaction)

Reagent	Volume per singleplex reaction (μL)	Volume per multiplex reaction (μL)
ddH_2O	11.7	10.9
5× Green GoTaq flexi buffer	4	4
25 mM $MgCl_2$	2	2
10 mM dNTP	0.4	0.4
10 μM primer (*see* Table 1)	0.4 each	0.4 each
GoTaq DNA Polymerase (5u/μL)	0.1	0.1
Template DNA (working solution)	1	1

6. Place the gel on a gel documentation system, select UV as the light source, and obtain a gel image.
7. The result is conclusive if all the controls (negative and positive) give the expected amplicons. Amplicon sizes of 425, 583, and 663 bp indicate the Fov4 MT, T, and MiT genotypes, respectively. An amplicon size of 396 bp indicates the Fov4 N genotype only if the genotype Fov4 test described in Subheading 3.4.2 is also positive. If there is no amplicon, the DNA preparation is not amenable to amplification, or the DNA is not from Fov4 or *Fusarium oxysporum*. For DNA extracted directly from a plant sample, a further possibility is that the quantity/concentration of fungal DNA is below the detection limit (*see* **Note 13**).

3.4.2 Detection of Fov4 by Singleplex PCR

Detection of Fov4 is achieved by singleplex PCR using the primer pair FovT-F and FovA-R (Table 1) with either the DNA working solution prepared from fungal cultures (20 ng/mL) or prepared directly from the infected plant tissues (200 ng/mL).

1. Proceed as described in **steps 1–6** of Subheading 3.4.1, except that the reactions are run as a singleplex PCR, and only one positive control using the fungal DNA from any one of the known Fov4 isolates should be included in **step 3**.
2. The result is conclusive if all the controls (negative and positive) give the expected amplicons. An amplicon size of 500 bp indicates an Fov4 genotype sample. No amplicon indicates that the DNA preparation is not amenable to amplification or is not from Fov4. For DNA extracted directly from a plant sample, a further possibility is that the quantity/concentration of fungal DNA is below the detection limit (*see* **Note 13**).

3.4.3 Detection and Genotype Determination of Fov4 by Singleplex PCR

Singleplex PCRs can also be utilized to detect and determine the Fov4 N genotype with the primer pair FovP-F and FovP-R, the Fov4 T or MiT genotype with the primer pair FovP-F and FovT-R, and the Fov4 MT genotype with the primer pair FovP-F and FovM-R. Threefold more PCR reactions are needed for singleplex PCR compared to the multiplex reactions. However, these singleplex PCR reactions may provide better amplification efficiency and specificity for certain samples than the multiplex reactions described in Subheading 3.4.1, because the former targets only one locus while the latter targets multiple loci in a single reaction. For DNA samples that failed to produce a positive amplicon due to the less than ideal quality of DNA, running these singleplex PCR may provide a better chance of producing a positive confirmation if the sample is of the Fov4 genotype.

1. For each N, T/MiT, or MT genotype determination, proceed as described in **steps 1–6** of Subheading 3.4.1, except that the

reactions are run as a singleplex PCR, and (1) only one positive control using the fungal DNA from a known Fov4 N genotype isolate should be included in **step 3** for the Fov N genotype determination; (2) two positive controls, one from an Fov4 T and one from an Fov4 MiT genotype isolate, are needed for the Fov4 T/MiT genotype determination; and (3) only one positive control from an Fov4 MT genotype isolate is needed for the Fov4 MT genotype determination.

2. The result is conclusive if all the controls (negative and positive) give the expected amplicons. Amplicon sizes of 425, 583, and 663 bp indicate the Fov4 MT, T, and MiT genotype, respectively. An amplicon size of 396 bp indicates the Fov4 N genotype only if the genotype Fov4 test described in Subheading 3.4.2 is also positive.

4 Notes

1. Other closely related *Fusarium* species, such as *F. fujikuroi*, are often recovered from wilted cotton samples, and their colony appearance resembles Fov [9, 10]. If the cotton samples were infected by *Verticillium dahliae*, another important wilt pathogen of cotton, these isolates can also be recovered from this medium [42]; however *V. dahliae* isolates grow slower than *F. oxysporum* isolates on this medium.
2. This will keep PCNB uniformly suspended in the medium.
3. If not used within a few days, bag the plates and store at 4 °C. Allow the plates to warm up to room temperature 1–2 h before using them.
4. Sample collection can be conducted at any cotton growth stage after seedling emergence. We were able to recover Fov4 from wilted cotton from late May to early August [10]. However, the infested area will be harder to spot as the canopy of neighboring plants covers the gaps created by stand loss or stunting.
5. Avoid picking dead plants as tissues from these samples will be inundated with secondary infections.
6. Pith darkening at the cut ends is a good indication of infection by Fov4. However, Fov4 had been recovered often from stems with no obvious brown staining. Also, vascular browning is caused by several different cotton wilt pathogens, and the vascular phenotype of various genotypes of Fov and *V. dahliae* infections often appears similar.
7. In case the samples cannot be processed immediately, they can be kept in brown paper bags and dried in a ventilated, cool, and dry area. Fov4 has been isolated by selective plating from samples that were stored this way for a few months.

8. Properly collected plant samples that had minimal secondary infections from Fov4 infested fields can also be plated on non-selective potato dextrose agar amended with chloramphenicol and tetracycline (PDAA), yielding mainly Fov4 colonies.
9. The pestles can be reused after decontamination by autoclaving or by soaking in fresh 1.5% hypochlorite for 30 min. Then the pestles are washed in soapy water, rinsed in tap water, distilled water, and autoclaved prior to the next use.
10. This is a convenient stopping point for the day if needed, and the processed tubes can be placed in a labeled rack in a −20 °C freezer.
11. Do not over dry the DNA, which will cause difficulty in redissolving the DNA pellet.
12. Use caution when handling ethidium bromide, a potential carcinogen. Use gloves for electrophoresis work and change gloves after the work. Properly dispose of ethidium bromide-contaminated waste materials.
13. For those samples that fail to produce an amplicon, the test is rerun with the primer pair EF-1A (5′- CACCT TAACGTCGTCGTCATC-3′) and EF516R (5′-AAGAACC CAGGCGTACTTGA-3′) as a control for DNA from fungal cultures. A 497 bp amplicon indicates the DNA preparation is suitable for PCR and is from a *F. oxysporum* isolate. For DNA extracted from plant tissues, the reaction is rerun with the primer pair CelA1-F (5′-GGATCTGCACCCATCAATCT-3′) and CelA1-R (5′-GCAAAGAGATGGGCTGAAAC-3′) [43] as a control. A 401 bp amplicon indicates the DNA preparation is suitable for PCR and is from cotton.

Acknowledgments

We thank Cotton Incorporated for their partial support of this research.

Disclaimer: The mention of firm names or trade products does not imply that they are endorsed or recommended by the US Department of Agriculture (USDA) over other firms or similar products not mentioned. The USDA is an equal opportunity provider and employer.

References

1. Armstrong GM, Armstrong JK (1960) American, Egyptian, and Indian cotton-wilt Fusaria: their pathogenicity and relationship to other wilt Fusaria. Tech Bull U S Dept Agric 1219:1–19
2. Chen Q, Ji X, Sun W (1985) Identification of races of cotton wilt *Fusarium* in China. Sci Agric Sin 6:1–6

3. Armstrong GM, Armstrong JK (1978) A new race (race 6) of the cotton-wilt *Fusarium* from Brazil. Plant Dis Rep 62(5):421–423
4. Ibrahim FM (1966) A new race of cotton-wilt *Fusarium* in the Sudan Gezira. Empire Cotton Growing Rev 43:296–299
5. Ortiz CS, Bell AA, Magill CW, Liu J (2017) Specific PCR detection of *Fusarium oxysporum* f. sp. *vasinfectum* California race 4 based on a unique *Tfo1* insertion event in the *PHO* gene. Plant Dis 101:34–44
6. Abo K, Klein KK, Edel-Hermann V, Gautheron N, Traore D, Steinberg C (2005) High genetic diversity among strains of *Fusarium oxysporum* f. sp *vasinfectum* from cotton in Ivory Coast. Phytopathology 95:1391–1396
7. Cianchetta AN, Allen TW, Hutmacher RB, Kemerait RC, Kirkpatrick TL, Lawrence GW, Lawrence KS, Mueller JD, Nichols RL, Olsen MW, Overstreet C, Woodward JE, Davis RM (2015) Survey of *Fusarium oxysporum* f. sp. *vasinfectum* in the United States. J Cotton Sci 19:328–336
8. Skovgaard K, Nirenberg HI, O'Donnell K, Rosendahl S (2001) Evolution of *Fusarium oxysporum* f. sp. *vasinfectum* races inferred from multigene genealogies. Phytopathology 91:1231–1237
9. Bell AA, Kemerait RC, Ortiz CS, Prom S, Quintana J, Nichols RL, Liu J (2017) Genetic diversity, virulence, and *Meloidogyne incognita* interactions of *Fusarium oxysporum* isolates causing cotton wilt in Georgia. Plant Dis 101:948–956
10. Bell AA, Gu A, Olvey J, Wagner TA, Tashpulatov JJ, Prom S, Quintana J, Nichols RL, Liu J (2019) Detection and characterization of *Fusarium oxysporum* f. sp. *vasinfectum* VCG0114 (race 4) isolates of diverse geographic origins. Plant Dis 103:1998–2009
11. Holmes EA, Bennett RS, Spurgeon DW, Colyer PD, Davis RM (2009) New genotypes of *Fusarium oxysporum* f. sp. *vasinfectum* from the southeastern United States. Plant Dis 93:1298–1304
12. Davis RD, Moore NY, Kochman JK (1996) Characterisation of a population of *Fusarium oxysporum* f. sp. *vasinfectum* causing wilt of cotton in Australia. Aust J Agric Res 47:1143–1156
13. Kim Y, Hutmacher RB, Davis RM (2005) Characterization of California isolates of *Fusarium oxysporum* f. sp. *vasinfectum*. Plant Dis 89:366–372
14. Liu J, Bell AA, Wheeler MH, Stipanovic RD, Puckhaber LS (2011) Phylogeny and pathogenicity of *Fusarium oxysporum* isolates from cottonseed imported from Australia into California for dairy cattle feed. Can J Microbiol 57:874–886
15. Leslie JF (1993) Fungal vegetative compatibility. Annu Rev Phytopathol 31:127–150
16. Fernandez D, Assigbetse K, Dubois MP, Geiger JP (1994) Molecular characterization of races and vegetative compatibility groups in *Fusarium oxysporum* f. sp. *vasinfectum*. Appl Environ Microbiol 60:4039–4046
17. Katan J, Fishler G, Grinstein A (1983) Short-term and long-term effects of soil solarization and crop sequence on Fusarium wilt and yield of cotton in Israel. Phytopathology 73:1215–1219
18. Bell AA, Liu J, Ortiz CS, Quintana J, Stipanovic RD, Crutcher FK (2016) Population structure and dynamics among *Fusarium oxysporum* isolates causing wilt of cotton. In: Boyd S, Huffman M (eds) Proceedings of beltwide cotton conferences, New Orleans, LA. National Cotton Council of America, Memphis, TN, pp 153–158
19. Garber RH, Jorgenson EC, Smith S, Hyer AH (1979) Interaction of population-levels of *Fusarium oxysporum* f. sp. *vasinfectum* and *Meloidogyne incognita* on cotton. J Nematol 11:133–137
20. Jorgenson EC, Hyer AH, Garber RH, Smith SN (1978) Influence of soil fumigation on *Fusarium*-root-knot nematode disease complex of cotton in California. J Nematol 10:228–231
21. da Silva MB, Davis RF, Doan HK, Nichols RL, Kemerait RC, Halpern HC, Brewer MT, Jagdale G, Chee PW (2019) Fusarium wilt of cotton may commonly result from the interaction of *Fusarium oxysporum* f. sp. *vasinfectum* with *Belonolaimus longicaudatus*. J Nematol 51:1–10
22. Wang C, Roberts PA (2006) A Fusarium wilt resistance gene in *Gossypium barbadense* and its effect on root-knot nematode-wilt disease complex. Phytopathology 96:727–734
23. Kappelman AJ (1983) Distribution of races of *Fusarium oxysporum* f. sp. *vasinfectum* within the United States. Plant Dis 67:1229–1231
24. Bennett RS, Scott TZ, Lawrence KS, Lawrence GW (2013) Sequence characterization of race 4-like isolates of *Fusarium oxysporum* from Alabama and Mississippi. J Cotton Sci 17:125–130
25. Hutmacher RB, Ulloa M, Wright SD, Campbell BT, Percy R, Wallace T, Myers G, Bourland F, Weaver D, Chee P, Thaxton P, Zhang JF, Smith W, Dever J, Kuraparthy V, Bowman D, Jones D, Burke J (2013) Elite

upland cotton germplasm-pool assessment of Fusarium wilt resistance in California. Agron J 105:1635–1644

26. Halpern HC, Bell AA, Wagner TA, Liu J, Nichols RL, Olvey J, Woodward JE, Sanogo S, Jones CA, Chan CT, Brewer MT (2018) First report of Fusarium wilt of cotton caused by *Fusarium oxysporum* f. sp. *vasinfectum* race 4 in Texas, U.S.A. Plant Dis 102:446–446
27. Zhu Y, Lujan PA, Wedegaertner T, Nichols R, Abdelraheem A, Zhang JF, Sanogo S (2020) First report of *Fusarium oxysporum* f. sp. *vasinfectum* race 4 causing Fusarium wilt of cotton in New Mexico, U.S.A. Plant Dis 102:588–588
28. Davis RM, Colyer PD, Rothrock CS, Kochman JK (2006) Fusarium wilt of cotton: population diversity and implication for management. Plant Dis 90:692–703
29. Taubenhaus JJ, Ezekiel WN (1932) Seed transmission of cotton wilt. Science 76:61–62
30. Taubenhaus JJ, Christenson LD (1936) Role of insects in the distribution of cotton wilt caused by *Fusarium vasinfectum*. J Agric Res 53:0703–0712
31. Bennett RS, Hutmacher RB, Davis RM (2008) Seed transmission of *Fusarium oxysporum* f. sp. *vasinfectum* race 4 in California. J Cotton Sci 12:160–164
32. Smith SN, Snyder WC (1975) Persistence of *Fusarium oxysporum* f. sp. *vasinfectum* in fields in absence of cotton. Phytopathology 65:190–196
33. Yang ME, Davis RM, Hutmacher RB (2006) Fusarium wilt of cotton in California: characterization and PCR-based detection of race 4. In: Boyd S, Huffman M, Robertson B (eds) Beltwide cotton conferences, San Antonio, Texas, 2006. The Cotton Foundation, Cordova, TN, pp 93–96
34. Doan HK, Zhang S, Davis RM (2014) Development and evaluation of AmplifyRP Acceler8 diagnostic assay for the detection of *Fusarium oxysporum* f. sp. *vasinfectum* race 4 in cotton. Plant Health Progress 15(1):48–52
35. Crutcher FK, Doan HK, Bell AA, Davis RM, Stipanovic RD, Nichols RL, Liu J (2016) Evaluation of methods to detect the cotton wilt pathogen *Fusarium oxysporum* f. sp. *vasinfectum* race 4. Eur J Plant Pathol 144:225–230
36. Zhang JF, Zhu Y, Abdelraheem A, Lujan P, Idowu J, Nichols R, Wedegaertner T (2020) Field survey, detection and characterization of Fusarium wilt race 4 in cotton in New Mexico. In: Proceedings of the beltwide cotton conference. National Cotton Council of America, Memphis, TN, pp 84–84
37. Diaz J, Garcia J, Lara C, Hutmacher RB, Ulloa M, Nichols RL, Ellis ML (2020) Characterization of current *Fusarium oxysporum* f. sp. *vasinfectum* isolates from cotton in the San Joaquin Valley of California and Lower Valley El Paso Texas. Plant Dis. https://doi.org/10.1094/pdis-05-20-1038-re
38. Komada H (1975) Development of a selective medium for quantitative isolation of *Fusarium oxysporum* from natural soil. Rev Plant Protect Res 8:114–124
39. Nelson PE, Toussoun TA, Marasas WFO (1983) *Fusarium* species: an illustrated manual for identification. Pennsylvania State University Press, University Park, PA
40. Woloshuk CP, Seip ER, Payne GA, Adkins CR (1989) Genetic transformation system for the aflatoxin-producing fungus *Aspergillus flavus*. Appl Environ Microbiol 55:86–90
41. Li H, Luo J, Hemphill JK, Wang JT, Gould JH (2001) A rapid and high yielding DNA miniprep for cotton (*Gossipium spp.*). Plant Mol Biol Rep 19:183a–183e
42. Wagner T, Gu A, Duke SE, Bell AA, Magill CW, Liu J (2020) Genetic diversity and pathogenicity of *Verticillium dahliae* isolates and their co-occurrence with *Fusarium oxysporum* f. sp. *vasinfectum* causing cotton wilt in Xinjiang, China. Plant Dis 105(4):978–985. https://doi.org/10.1094/pdis-09-20-2038-re
43. Xin ZG, Velten JP, Oliver MJ, Burke JJ (2003) High-throughput DNA extraction method suitable for PCR. BioTechniques 34:820–826

Index

Jeffrey Coleman (ed.), *Fusarium wilt: Methods and Protocols*,
Methods in Molecular Biology, vol. 2391, https://doi.org/10.1007/978-1-0716-1795-3,

MIX
Papier aus verantwortungsvollen Quellen
Paper from responsible sources
FSC® C105338

If you have any concerns about our products,
you can contact us on
ProductSafety@springernature.com

In case Publisher is established outside the EU,
the EU authorized representative is:
Springer Nature Customer Service Center GmbH
Europaplatz 3, 69115 Heidelberg, Germany

Printed by Libri Plureos GmbH
in Hamburg, Germany